FRACTURE MECHANICS

Fracture and "slow" crack growth reflect the response of a material (i.e., its microstructure) to the conjoint actions of mechanical and chemical driving forces and are affected by temperature. Therefore, there is a need for quantitative understanding and modeling of the influences of chemical and thermal environments, and of microstructure, in terms of the key internal and external variables and for their incorporation into design and probabilistic implications. This text, which the author has used in a fracture mechanics course for advanced undergraduate and graduate students, is based on the work of the author's Lehigh University team whose integrative research combined fracture mechanics, surface and electrochemistry, materials science, and probability and statistics to address a range of fracture safety and durability issues on aluminum, ferrous, nickel, and titanium alloys, and ceramics. Examples from this research are included to highlight the approach and applicability of the findings in practical durability and reliability problems.

Robert P. Wei is the Reinhold Professor of Mechanical Engineering and Mechanics at Lehigh University. His principal research is in fracture mechanics, including chemical, microstructural, and mechanical considerations of stress corrosion cracking, fatigue, and corrosion, and in life-cycle engineering. He is the author of hundreds of refereed research publications. He is a Fellow of the American Society for Testing and Materials; the American Society of Metals International; and the American Institute of Mining, Metallurgical, and Petroleum Engineering and a member of Sigma Xi and the Phi Beta Delta International Honor Societies.

Fracture Mechanics

INTEGRATION OF MECHANICS, MATERIALS SCIENCE, AND CHEMISTRY

Robert P. Wei

Lehigh University

Shaftesbury Road, Cambridge CB2 8EA, United Kingdom

One Liberty Plaza, 20th Floor, New York, NY 10006, USA

477 Williamstown Road, Port Melbourne, VIC 3207, Australia

314–321, 3rd Floor, Plot 3, Splendor Forum, Jasola District Centre, New Delhi – 110025, India

103 Penang Road, #05–06/07, Visioncrest Commercial, Singapore 238467

Cambridge University Press is part of Cambridge University Press & Assessment, a department of the University of Cambridge.

We share the University's mission to contribute to society through the pursuit of education, learning and research at the highest international levels of excellence.

www.cambridge.org
Information on this title: www.cambridge.org/9780521194891

First published 2010
First paperback edition 2013

A catalogue record for this publication is available from the British Library

Library of Congress Cataloging-in-Publication data
Wei, Robert Peh-ying, 1931–
Fracture mechanics : integration of mechanics, materials science, and chemistry / Robert Wei.
    p. cm.
Includes bibliographical references.
ISBN 978-0-521-19489-1 (hardback)
1. Fracture mechanics.   I. Title.
TA409.W45   2010
620.1′126–dc22   2009044098

ISBN   978-0-521-19489-1   Hardback
ISBN   978-1-107-66552-1   Paperback

To Lee

For her love, counsel, dedication, and support

Contents

Preface

Engineering Fracture Mechanics, as a recognized branch of engineering mechanics, had its beginning in the late 1940s and early 1950s, and experienced major growth through the next three decades. The initial efforts were driven primarily by naval and aerospace interests. By the end of the 1980s, most of the readily tractable mechanics problems had been solved, and computational methods have become the norm in solving practical problems in fracture/structural integrity. On the lifing ("slow" crack growth) side, the predominant emphasis has been on empirical characterization and usage of data for life prediction and reliability assessments.

In reality, fracture and "slow" crack growth reflect the response of a material (*i.e.*, its microstructure) to the conjoint actions of mechanical and chemical driving forces, and are affected by temperature. The need for quantitative understanding and modeling of the influences of chemical and thermal environments and of microstructure (*i.e.*, in terms of the key *internal* and *external* variables), and for their incorporation into design, along with their probabilistic implications, began to be recognized in the mid-1960s.

With support from AFOSR, ALCOA, DARPA, DOE (Basic Energy Sciences), FAA, NSF, ONR, and others, from 1966 to 2008, the group at Lehigh University undertook integrative research that combined fracture mechanics, surface and electrochemistry, materials science, and probability and statistics to address a range of fracture safety and durability issues on aluminum, ferrous, nickel, and titanium alloys and on ceramics. Examples from this research are included to highlight the approach and applicability of the findings in practical problems of durability and reliability. An appended list of publications provides references/sources for more detailed information on research from the overall program.

The title *Fracture Mechanics: Integration of Fracture Mechanics, Materials Science, and Chemistry* gives tribute to those who have shared the vision and have contributed to and supported this long-term, integrative effort, and to those who recognize the need and value for this multidisciplinary team effort.

The author has used the material in this book in a fracture mechanics course for advanced undergraduate and graduate students at Lehigh University. This book should also serve as a reference for the design and management of engineered systems.

Acknowledgments

The author acknowledges the invaluable contributions and dedication of his colleagues: Dr. Ye T. (Russell) Chou (Materials Science), Dr. Kamil Klier (Surface Chemistry), Dr. Gary Simmons (Surface Chemistry), Dr. D. Gary Harlow (Probability and Statistics/Mechanical Engineering & Mechanics), and Dr. Ming Gao (Materials Science), and the many postdoctoral researchers and graduate students in Mechanical Engineering and Mechanics, Materials Science and Engineering, and Surface Science and Electrochemistry, who made this possible. The author also acknowledges the International Multimedia Resource Center (IMRC) of Lehigh University, under the leadership of Johanna Brams, especially Nyko DePeyer and Dawn Dayawon, for their assistance in graphic arts and manuscript preparation, and Sharon Siegler, Lehigh University librarian, for her counsel and expert assistance.

1   Introduction

Fracture mechanics, or the mechanics of fracture, is a branch of engineering science that addresses the problem of the integrity and durability of materials or structural members containing cracks or cracklike defects. The presence of cracks may be real, having been introduced through the manufacturing processes or during service. On the other hand, their presence may have to be assumed because limitations in the sensitivity of nondestructive inspection procedures preclude full assurance of their absence. A perspective view of fracture mechanics can be gained from the following questions:

- How much load will it carry, with and without cracks? (a question of *structural safety and integrity*).
- How long will it last, with and without cracks? Alternatively, how much longer will it last? (a concern for *durability*).
- Are you sure? (the important issue of *reliability*).
- How sure? (*confidence* level).

The corollary questions are as follows, and will not be addressed here:

- How much will it cost? To buy? (capital or acquisition cost); to run? (operational cost); to get rid of? (disposal/recycling cost)
- Optimize capital (acquisition) costs?
- Optimize overall (life cycle) cost?

These questions appear to be simple, but are in fact profound and difficult to answer. Fracture mechanics attempts to address (or provides the framework for addressing) these questions, where the presence of a crack or cracklike defects is presumed.

The first of the questions deals with the stability of a crack under load. Namely, would it remain stable or grow catastrophically? The second question deals with the issue: "if a crack can grow stably under load, how long would it take before it reaches a length to become unstable, or become unsafe?" The third question, encompassing the first two, has to do with certainty; and the last deals with the confidence in the answers. These questions lead immediately to other questions.

Can the properties that govern crack stability and growth be computed on the basis of first principles, or must they be determined experimentally? How are these properties to be defined, and how well can they be determined? What are the variations in these properties? If the failure load or crack growth life of a material can be measured, what degree of certainty can be attached to the prediction of safe operating load or serviceable life of a structural component made from that material?

1.1 Contextual Framework

In-service incidents provide lasting reminders of the "aging" of, or cracking in, engineered systems. Figure 1.1 shows the consequence of an in-flight rupture of an eighteen-foot section of the fuselage of an Aloha Airlines 737 aircraft over the Hawaiian Islands in 1988. The rupture was attributed to the "link up" of extensive fatigue cracking along a riveted longitudinal joint. Fortunately, the pilots were

Figure 1.1. In-flight separation of an upper section of the fuselage of a B737-200 aircraft in 1988 attributed to corrosion and fatigue.

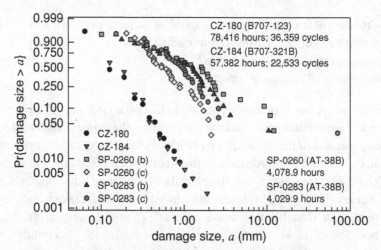

Figure 1.2. Damage distribution in aged B707 (CZ-180 and CZ184) after more than twenty years of service, and AT-38B aircraft after more than 4,000 hours of service [3].

able to land the aircraft safely, with the loss of only one flight attendant who was serving in the cabin. Tear-down inspection data on retired commercial transport and military aircraft [1, 2] (Fig. 1.2), provide some sense of the damage that can accrue in engineered structures, and of the need for robust design, inspection, and maintenance.

On the other end of the spectrum, so to speak, the author encountered a fatigue failure in the "Agraph" of a chamber grand piano (Figs. 1.3 and 1.4). An Agraph is typically a bronze piece that supports the keyboard end of piano strings (wires). It

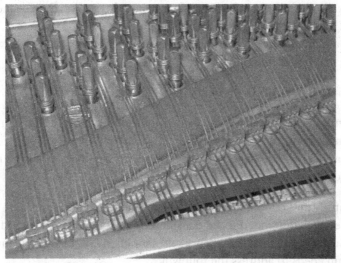

Figure 1.3. Interior of a chamber grand piano showing a row of Agraphs aligned just in front of the red velvet cushion.

Figure 1.4. (left) Photograph of a new Agraph from a chamber grand piano, and (right) scanning electron micrographs of the mating halves of a fractured Agraph showing fatigue markings and final fracture.

sets the effective length of the strings and carries the effect of tension in the strings that ensures proper tuning. As such, it carries substantial static (from tuning tension) and vibratory loads (when the string is struck) and undergoes fatigue.

1.2 Lessons Learned and Contextual Framework

Key lessons learned from aging aircraft and other research over the past four decades showed that:

- Empirically based, discipline-specific methodologies for design and management of engineered systems are not adequate.
- Design and management methodologies need to be science-based, much more holistic, and better integrated.

Tear-down inspections of B-707 and AT-38B aircraft [1, 2] showed:

- The significance of localized corrosion on the evolution and distribution of fatigue damage was not fully appreciated.
- Its impact could not have been predicted by the then existing and current technologies.

As such, transformation in thinking and approach is needed.

Fracture mechanics need to be considered in the context of a modern design paradigm. Such a contextual framework and simplified flow chart is given in Fig. 1.5. The paradigm needs to address the following:

- Optimization of life-cycle cost (*i.e.*, cost of ownership)
- System/structural integrity, performance, safety, durability, reliability, etc.
- Enterprise planning
- Societal issues (*e.g.*, environmental impact)

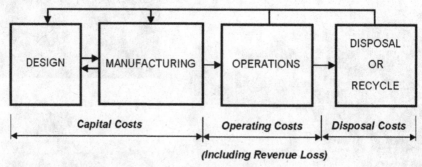

Figure 1.5. Contextual framework and simplified flow diagram for the design and management of engineered systems.

A schematic flow diagram that underlies the processes of reliability and safety assessments is depicted in Fig. 1.6. The results should be used at different levels to aid in operational and strategic planning.

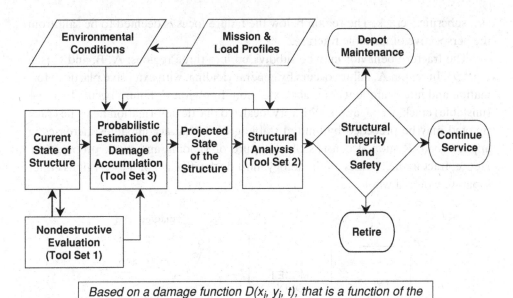

Figure 1.6. Simplified flow diagram for life prediction, reliability assessment, and management of engineered systems.

Fracture mechanics, therefore, must deal with the following two classes of problems:

- Crack tolerance or residual strength
- Crack growth resistance

A brief consideration of each is given here to identify the nature of the problems, and to assist in defining the scope of the book.

1.3 Crack Tolerance and Residual Strength

The concept of crack tolerance and residual strength can be understood by considering the fracture behavior of a plate, containing a central crack of length $2a$, loaded in remote tension under uniform stress σ (see Fig. 1.7). The fracture behavior is illustrated schematically also in Fig. 1.7 as a plot of failure stress versus half-crack length (a). The line drawn through the data points represents the failure locus, and the stress levels corresponding to the uniaxial yield and tensile strengths are also indicated. The position of the failure locus is a measure of the material's crack tolerance, with greater tolerance represented by a translation of the failure locus to longer crack lengths (or to the right).

The stress level corresponding to a given crack length on the failure locus is the residual strength of the material at that crack length. The residual strength typically would be less than the uniaxial yield strength. The crack length corresponding to a given stress level on the failure locus is defined as the critical crack size. A crack that is smaller (shorter) than the critical size, at the corresponding stress level, is defined as a subcritical crack. The region below the failure locus is deemed to be safe from the perspective of unstable fracture.

The fracture behavior may be subdivided into three regions: A, B, and C (see Fig. 1.7). In region A, failure occurs by general yielding, with extensive plastic deformation and minor amounts of crack extension. In region C, failure occurs by rapid (unstable) crack propagation, with very localized plastic deformation near the crack tip, and may be preceded by limited stable growth that accompanies increases in applied load. Region B consists of a mixture of yielding and crack propagation. Hence, fracture mechanics methodology must deal with each of these regions either separately or as a whole.

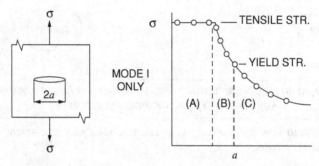

Figure 1.7. Schematic illustration of the fracture behavior of a centrally cracked plate loaded in uniform remote tension.

In presenting Fig. 1.7, potential changes in properties with time and loading rate and other time-dependent behavior were not considered. In effect, the failure locus should be represented as a surface in the stress, crack size, and time (or strain rate, or crack velocity) space (see Fig. 1.8). The crack tolerance can be degraded because of the strain rate sensitivity of the material, and time-dependent changes in microstructure (e.g., from strain aging and radiation damage), with concomitant increases in strength. As a result, even without crack extension and increases in applied load (or stress), conditions for catastrophic failure may be attained with time or an increase in applied load (or stress), or an increase in loading rate (see path 3 in Fig. 1.8b).

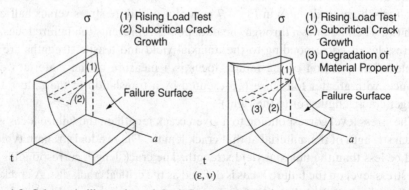

Figure 1.8. Schematic illustration of the influence of time (or strain rate, or crack velocity) on the fracture behavior of a centrally cracked plate loaded in uniform remote tension.

1.4 Crack Growth Resistance and Subcritical Crack Growth

Under certain loading (such as fatigue) and environmental (both internal and external to the material) conditions, cracks can and do grow and lead to catastrophic failure. The path for such an occurrence is illustrated by path 2 in Fig. 1.8. Because the crack size remains below the critical size during its growth, the processes are broadly termed subcritical crack growth. The rate of growth is determined by some appropriate driving force and growth resistance, which both must be defined by fracture mechanics.

The phenomenon of subcritical crack growth may be subdivided into four categories according to the type of loading and the nature of the external environment as shown in Table 1.1.

Table 1.1. *Categories of subcritical crack growth*

Loading condition	Inert environment	Deleterious environment
Static or sustained	Creep crack growth (or internal embrittlement)	Stress corrosion cracking
Cyclic or fatigue	Mechanical fatigue	Corrosion fatigue

Under statically applied loads, or sustained loading, in an inert environment, crack growth is expected to result from localized deformation near the crack tip. This phenomenon is of particular importance at elevated temperatures. Under cyclically varying loads, or in fatigue, crack growth can readily occur by localized, but reversed deformation in the crack-tip region. When the processes are assisted by the presence of an external, deleterious environment, crack growth is enhanced and is termed environmentally assisted crack growth.

Environmentally enhanced crack growth is typically separated into stress corrosion cracking (for sustained loading) and corrosion fatigue (for cyclic loading), and involves complex interactions among the environment, microstructure, and applied loading. Crack growth can occur also because of embrittlement by dissolved species (such as hydrogen) in the microstructure. This latter problem may be viewed in combination with deformation-controlled growth, or as a part of environmentally assisted crack growth.

1.5 Objective and Scope of Book

The objective of this book is to demonstrate the need for, and the efficacy of, a mechanistically based probability approach for addressing the structural integrity, durability, and reliability of engineered systems and structures. The basic elements of engineering fracture mechanics, materials science, surface and electrochemistry, and probability and statistics that are needed for the understanding of materials behavior and for the application of fracture mechanics-based methodology in design and research are summarized. Through examples used in this book, the need for and efficacy of an integrated, multidisciplinary approach is demonstrated.

The book is topically divided into four sections. In Chapters 2 and 3, the physical basis of fracture mechanics and the stress analysis of cracks, based on linear elasticity, are summarized. In Chapters 4 and 5, the experimental determination of fracture toughness and the use of this property in design are highlighted (How much load can be carried?). Chapters 6 to 9 address the issue of durability (How long would it last?), and cover the interactions of mechanical, chemical, and thermal environments. Selected examples are used to illustrate the different cracking response of different material/environment combinations, and the influences of temperature, loading frequency, etc. The development of mechanistic understanding and modeling is an essential outcome of these studies. Chapter 10 illustrates the use of the forgoing mechanistically based models in the formulation of probability models in quantitative assessment of structural reliability and safety. It serves to demonstrate the need to transition away from the traditional empirically based design approaches, and the attendant uncertainties in their use in structural integrity, durability, and reliability assessments.

The book (along with the appended list of references) serves as a reference source for practicing engineers and scientists, in engineering, materials science, and chemistry, and as a basis for the formation of multidisciplinary teams. It may be used as a textbook for seniors and graduate students in civil and mechanical engineering, and materials science and engineering, and as a basis for the formation of multidisciplinary teams in industry and government laboratories.

REFERENCES

[1] Hug, A. J., "Laboratory Inspection of Wing Lower Surface Structure from 707 Aircraft for the J-STARS Program," The Boeing Co., FSCM81205, Document D500-12947-1, Wichita, KS, April 1994 (1996).

[2] Kimball, C. E., and Benac, D. J., "Analytical Condition Inspection (ACI) of AT-38B Wings," Southwest Research Institute, Project 06-8259, San Antonio, TX (1997).

[3] Harlow, D. G., and Wei, R. P., "Probability Modeling and Statistical Analysis of Damage in the Lower Wing Skins of Two Retired B-707 Aircraft," *Fatigue and Fracture of Engineering Materials and Structures*, 24 (2001), 523–535.

2 Physical Basis of Fracture Mechanics

In this chapter, the classical theories of failure are summarized first, and their inadequacy in accounting for the failure (fracture) of bodies that contain crack(s) is highlighted. The basic development of fracture mechanics, following the concept first formulated by A. A. Griffith [1, 2], is introduced. The concepts of strain energy release rate and stress intensity factor, and their identification as the *driving force* for crack growth are introduced. The experimental determinations of these factors are discussed. Fracture behavior of engineering materials is described, and the importance of fracture mechanics in the design and sustainment of engineered systems is considered.

2.1 Classical Theories of Failure

Classical theories of failure are based on concepts of maximum stress, strain, or strain energy and assume that the material is homogeneous and free from defects. Stresses, strains, and strain energies are typically obtained through elastic analyses.

2.1.1 Maximum Principal Stress (or Tresca [3]) Criterion

The *maximum principal stress criterion* for failure simply states that failure (by yielding or by fracture) would occur when the maximum principal stress reaches a critical value (*i.e.*, the material's yield strength, σ_{YS}, or fracture strength, σ_f, or tensile strength, σ_{UTS}). For a three-dimensional state of stress, given in terms of the Cartesian coordinates x, y, and z in Fig. 2.1 and represented by the left-hand matrix in Eqn. (2.1), a set of principal stresses (see Fig. 2.1) can be readily obtained by transformation:

$$
\begin{vmatrix} \sigma_{xx} & \tau_{xy} & \tau_{xz} \\ \tau_{yx} & \sigma_{yy} & \tau_{yz} \\ \tau_{zx} & \tau_{zy} & \sigma_{zz} \end{vmatrix} \Rightarrow \begin{vmatrix} \sigma_1 & 0 & 0 \\ 0 & \sigma_2 & 0 \\ 0 & 0 & \sigma_3 \end{vmatrix} \tag{2.1}
$$

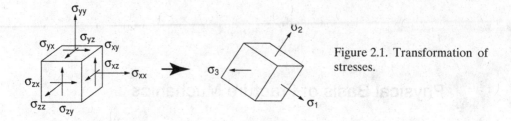

Figure 2.1. Transformation of stresses.

Assume that the largest principal stress is σ_1, the failure criterion is then given by Eqn. (2.2).

$$\sigma_1 = \sigma_{FAILURE} \ (\sigma_{YS} \text{ or } \sigma_f \text{ or } \sigma_{UTS}); \ \sigma_1 > \overset{\bullet}{\sigma}_2 > \sigma_3 \tag{2.2}$$

It is recognized that failure can also occur under compression. In that case, the strength properties in Eqn. (2.2) need to be replaced by the suitable ones for compression.

2.1.2 Maximum Shearing Stress Criterion

The *maximum shearing stress criterion* for failure simply states that failure (by yielding) would occur when the maximum shearing stress reaches a critical value (*i.e.*, the material's yield strength in shear). Taking the maximum and minimum principal stresses to be σ_1 and σ_3, respectively, then the failure criterion is given by Eqn. (2.3), where the yield strength in shear is taken to be one-half that for uniaxial tension.

$$\tau_{max} = \tau_c = \frac{(\sigma_1 - \sigma_3)}{2} \ \Rightarrow \ \frac{\sigma_{YS}}{2} \text{ for uniaxial tension} \tag{2.3}$$

2.1.3 Maximum Principal Strain Criterion

The *maximum principal strain criterion* for failure simply states that failure (by yielding or by fracture) would occur when the maximum principal strain reaches a critical value (*i.e.*, the material's yield strain or fracture strain, ε_f). Again taking the maximum principal strain (corresponding to the maximum principal stress) to be ε_1, the failure criterion is then given by Eqn. (2.4).

$$\varepsilon_1 = \varepsilon_{FAILURE} \ \Rightarrow \ \frac{\sigma_{YS}}{E} \text{ or } \varepsilon_f \text{ for uniaxial tension} \tag{2.4}$$

2.1.4 Maximum Total Strain Energy Criterion

The *total strain energy criterion* for failure states that failure (by yielding or by fracture) would occur when the total strain energy, or total strain energy density u_T, reaches a critical value u_c. The total strain energy density may be expressed in terms of the stresses and strains in the Cartesian coordinates, or the principal stresses and strains, by Eqn. (2.5).

$$u_T = \frac{1}{2}(\sigma_{xx}\varepsilon_{xx} + \sigma_{yy}\varepsilon_{yy} + \sigma_{zz}\varepsilon_{zz} + \tau_{xy}\gamma_{xy} + \tau_{yz}\gamma_{yz} + \tau_{zx}\gamma_{zx})$$

$$= \frac{1}{2E}\left[\sigma_{xx}^2 + \sigma_{yy}^2 + \sigma_{zz}^2 - 2v(\sigma_{xx}\sigma_{yy} + \sigma_{yy}\sigma_{zz} + \sigma_{zz}\sigma_{xx})\right] + \frac{1+v}{E}\left(\tau_{xy}^2 + \tau_{yz}^2 + \tau_{zx}^2\right)$$

$$u_T = \frac{1}{2}(\sigma_1\varepsilon_1 + \sigma_2\varepsilon_2 + \sigma_3\varepsilon_3) \tag{2.5}$$

$$= \frac{1}{2E}\left[\sigma_1^2 + \sigma_2^2 + \sigma_3^2 - 2v(\sigma_1\sigma_2 + \sigma_2\sigma_3 + \sigma_3\sigma_1)\right].$$

Failure occurs when $u_T = u_c$; or when

$$u_T \Rightarrow \frac{1}{2}\frac{\sigma_{YS}^2}{E} \quad \text{or} \quad \frac{1}{2}\frac{\sigma_f^2}{E} \quad \text{for uniaxial tension} \tag{2.6}$$

2.1.5 Maximum Distortion Energy Criterion

The total strain energy density may be subdivided into two parts; namely, *dilatation* and *distortion*, where dilatation is associated with changes in volume and distortion is associated with changes in shape that result from straining. In other words, $u_T = u_v + u_d$, or $u_d = u_T - u_v$. From Eqn. (2.5), the total strain energy density is given by:

$$u_T = \frac{1}{2E}\left[\sigma_{xx}^2 + \sigma_{yy}^2 + \sigma_{zz}^2 - 2v(\sigma_{xx}\sigma_{yy} + \sigma_{yy}\sigma_{zz} + \sigma_{zz}\sigma_{xx})\right] + \frac{1+v}{E}\left(\tau_{xy}^2 + \tau_{yz}^2 + \tau_{zx}^2\right)$$

$$= \frac{1}{2E}\left[\sigma_1^2 + \sigma_2^2 + \sigma_3^2 - 2v(\sigma_1\sigma_2 + \sigma_2\sigma_3 + \sigma_3\sigma_1)\right]$$

The strain energy density for dilatation (u_v) is given in terms of the hydrostatic stress:

$$u_v = \frac{1-2v}{6E}(\sigma_{xx} + \sigma_{yy} + \sigma_{zz})^2 = \frac{1-2v}{6E}(\sigma_1 + \sigma_2 + \sigma_3)^2$$

The distortion energy density and the *maximum distortion energy criterion for failure*, in terms of yielding, are given, therefore, by Eqns. (2.7) and (2.8).

$$u_d = \frac{1+v}{6E}\left[(\sigma_{xx} - \sigma_{yy})^2 + (\sigma_{yy} - \sigma_{zz})^2 + (\sigma_{zz} - \sigma_{xx})^2\right] + \frac{1+v}{E}\left[\tau_{xy}^2 + \tau_{yz}^2 + \tau_{zx}^2\right]$$

$$= \frac{1+v}{6E}\left[(\sigma_1 - \sigma_2)^2 + (\sigma_2 - \sigma_3)^2 + (\sigma_3 - \sigma_1)^2\right] \Rightarrow \frac{1+v}{3E}\sigma_{YS}^2 \tag{2.7}$$

or

$$\left[(\sigma_{xx} - \sigma_{yy})^2 + (\sigma_{yy} - \sigma_{zz})^2 + (\sigma_{zz} - \sigma_{xx})^2\right] + 6\left[\tau_{xy}^2 + \tau_{yz}^2 + \tau_{zx}^2\right]$$

$$= \left[(\sigma_1 - \sigma_2)^2 + (\sigma_2 - \sigma_3)^2 + (\sigma_3 - \sigma_1)^2\right] = 2k^2 = 2\sigma_{YS}^2 \tag{2.8}$$

2.1.6 Maximum Octahedral Shearing Stress Criterion (von Mises [4] Criterion)

This failure criterion is given in terms of the *octahedral shearing stress*. It is identical to the maximum distortion energy criterion, except that it is expressed in stress versus energy units. The criterion, expressed in terms of the principal stresses, is given in Eqn. (2.9).

$$\{[(\sigma_{xx} - \sigma_{yy})^2 + (\sigma_{yy} - \sigma_{zz})^2 + (\sigma_{zz} - \sigma_{xx})^2] + 6[\tau_{xy}^2 + \tau_{yz}^2 + \tau_{zx}^2]\}^{\frac{1}{2}}$$
$$= [(\sigma_1 - \sigma_2)^2 + (\sigma_2 - \sigma_3)^2 + (\sigma_3 - \sigma_1)^2]^{\frac{1}{2}} = \sqrt{2}k = \sqrt{2}\sigma_{YS} \qquad (2.9)$$

2.1.7 Comments on the Classical Theories of Failure

Criteria 2, 5, and 6 are generally used for yielding, or the onset of plastic deformation, whereas criteria 1, 3, and 4 are used for fracture. The maximum shearing stress (or Tresca [3]) criterion is generally not true for multiaxial loading, but is widely used because of its simplicity. The distortion energy and octahedral shearing stress criteria (or von Mises criterion [4]) have been found to be more accurate. None of the failure criteria works very well. Their inadequacy is attributed, in part, to the presence of cracks, and of their dominance, in the failure process.

2.2 Further Considerations of Classical Theories

It is worthwhile to consider whether the classical theories (or criteria) of failure can still be applied if the stress (or strain) concentration effects of geometric discontinuities (*e.g.*, notches and cracks) are properly taken into account. In other words, one might define a (theoretical) stress concentration factor, for example, to account for the elevation of local stress by the geometric discontinuity in a material and still make use of the maximum principal stress criterion to "predict" its strength, or load-carrying capability.

To examine this possibility, the case of an infinitely large plate of uniform thickness that contains an elliptical notch with semi-major axis *a* and semi-minor axis *b* (Fig. 2.2) is considered. The plate is subjected to remote, uniform in-plane tensile stresses (σ) perpendicular to the major axis of the elliptical notch as shown. The

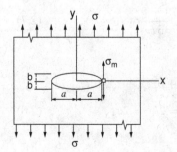

Figure 2.2. Schematic diagram of a plate, containing an elliptical notch, subjected to uniform, remote tension.

maximum tensile stress (σ_m) would occur at the ends of the major axis of the elliptical notch, and is given by the following relationship:

$$\sigma_m = \sigma \left(1 + \frac{2a}{b}\right) \tag{2.10}$$

The parenthetical term is the theoretical stress concentration factor for the notch. By squaring a/b and recognizing that b^2/a is the radius of curvature ρ, σ_m may be rewritten as follows:

$$\sigma_m = \sigma \left(1 + 2\sqrt{\frac{a^2}{b^2}}\right) = \sigma \left(1 + 2\sqrt{\frac{a}{\rho}}\right) \tag{2.11}$$

and

$$\sigma_m \approx 2\sigma \sqrt{\frac{a}{\rho}} \text{ for } \rho \ll a \tag{2.12}$$

As the root radius (or radius of curvature) approaches zero, or as the elliptical notch is collapsed to approximate a crack, then the maximum stress should approach infinity (*i.e.*, as $\rho \to 0$, $\sigma_m \to \infty$).

If the maximum principal stress criterion is to hold, then the ratio of the applied stress to cause fracture to the 'fracture stress' should approach zero as the radius of curvature is reduced to zero (*i.e.*, $\sigma/\sigma_f \to 0$ as $\rho \to 0$) in accordance with the following relationship:

$$\frac{\sigma}{\sigma_f} = \left(1 + 2\sqrt{\frac{a}{\rho}}\right)^{-1} \tag{2.13}$$

Comparisons with experimental data show that the stress required to produce fracture actually approached a constant (Fig. 2.3). Thus, the maximum principal stress criterion for failure, as well as the other classical criteria, is inadequate and inappropriate.

Further insight on fracture may be drawn from experimental work on the strength of glass fibers. The results indicated that the strength of a fiber depended on its length, with shorter fibers showing greater strengths. Its strength can be increased by polishing. Freshly made glass fibers were also found to be much stronger than those that have been handled (Fig. 2.4); with the fresh-fiber strength approaching the theoretical tensile strength of the order of one-tenth the elastic modulus

Figure 2.3. Schematic illustration of a comparison of predictions of Eqn. (2.13) with experimental observations.

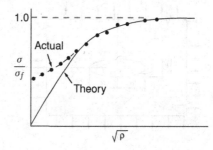

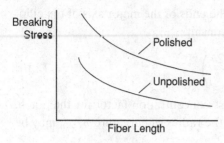

Figure 2.4. Schematic illustration of fracture strengths of polished and unpolished glass fibers.

(or E/10). These results suggested that the fiber strength was controlled by the presence of defects in the fibers. This suggestion was deduced from the following points:

1. The observed length dependence was consistent with the probabilistic considerations of defect distribution. The probability of encountering a defect being lower in a shorter fiber, therefore, could account for its greater strength.
2. The fact that polished fibers, and fresh fibers, were stronger suggested that the defects were predominately surface flaws (scratches, etc.), and confirmed the concept of defect-controlled fracture.

Thus, one needs a theory of fracture that is based on the stability of the largest (or dominant) flaw or crack in the material. Such formalism was first introduced by A. A. Griffith in 1920 [1] and forms the basis of what is now known as *linear (or linear elastic) fracture mechanics* (LEFM).

2.3 Griffith's Crack Theory of Fracture Strength

Griffith [1, 2] provided the first analysis of the equilibrium and stability of cracks in 1920 (paper first published in 1921; revised version published in 1924). He based his analysis on the consideration of the change in potential energy of a body into which a crack has been introduced. The equilibrium or stability of this crack under stress is then considered on the basis of energy balance. Griffith made use of the stress analysis results of Inglis [5] for a plate containing an elliptical notch and loaded in biaxial tension in computing the potential energy for deformation.

Consider, therefore, an infinitely large plate of elastic material of thickness B, containing a through-thickness crack of length $2a$, and subjected to uniform biaxial tension (σ) at infinity as shown in Fig. 2.5. Let U = potential energy of the system, U_o = potential energy of the system before introducing the crack, U_a = decrease

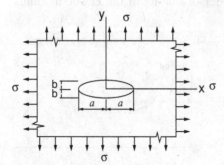

Figure 2.5. An infinitely large plate of elastic material containing a through-thickness central crack of length $2a$ and subjected to uniform biaxial tension σ.

in potential energy due to deformation (strain energy and boundary force work) associated with introduction of the crack, and U_γ = increase in surface energy due to the newly created crack surfaces. The potential energy of the system following the introduction of the crack then becomes:

$$U = U_o - U_a + U_\gamma \qquad (2.14)$$

Based on Inglis [5], the decrease in potential energy, for generalized plane stress, is given by:

$$U_a = \frac{\pi \sigma^2 a^2 B}{E} \qquad (2.15)$$

where E is the elastic (Young's) modulus. For plane strain, the numerator is modified by $(1 - v^2)$. For simplicity, however, this term will not be included in the subsequent discussions. The increase in surface energy (U_γ) is given by $4aB\gamma$, where γ is the surface energy (per unit area) and $4aB$ represents the area of the surfaces (each equals to $2aB$) created. Thus, the potential energy of the system becomes:

$$U = U_o - \frac{\pi \sigma^2 a^2 B}{E} + 4aB\gamma \qquad (2.16)$$

Since U_o is the potential energy of the system without a crack, it is therefore independent of the crack length a.

Equilibrium of the crack may be examined in terms of the variation in system potential energy with respect to crack length, a (with a minimum in potential energy constituting stable equilibrium, and a maximum, unstable equilibrium). Specifically,

$$\delta U = \frac{\partial U}{\partial a} \delta a = \left(-\frac{2\pi \sigma^2 a B}{E} + 4B\gamma \right) \delta a \qquad (2.17)$$

For maxima or minima, $\delta U = 0$. For a nonzero variation in a (or δa), then the expression inside the bracket must vanish; i.e.,

$$\frac{\pi \sigma^2 a}{E} = 2\gamma \qquad (2.18)$$

This is the equilibrium condition for a crack in an elastic, "brittle" material. Taking the second variation in U, one obtains:

$$\delta^2 U = \frac{\partial^2 U}{\partial a^2} \delta a = \left(-\frac{2\pi \sigma^2 B}{E} \right) \delta a < 0; \; (i.e., \text{always negative}) \qquad (2.19)$$

Therefore, the equilibrium is unstable.

The use of the concept of "equilibrium" in this context has been criticized by Sih and others. In more recent discussions of fracture mechanics, therefore, it is preferred to interpret the left-hand side of the equilibrium equation (2.18) as the generalized crack-driving force; i.e., the elastic energy per unit area of crack surface made available for an infinitesimal increment of crack extension, and is designated by G;

$$G = \frac{\pi \sigma^2 a}{E} \qquad (2.20)$$

The right-hand side is identified with the material's resistance to crack growth, R, in terms of the energy per unit area required in extending the crack ($R = 2\gamma$). Unstable fracturing would occur when the energy made available with crack extension (*i.e.*, the crack-driving force G) exceeds the work required (or R) for crack growth. The critical stress required to produce fracture (unstable or rapid crack growth) is then given by setting G equal to R:

$$\sigma_{cr} = \sqrt{\frac{2E\gamma}{\pi a}} \tag{2.21}$$

In other words, the critical stress for fracture σ_{cr} is inversely proportional to the square root of the crack size a.

Equation (2.21) may be rewritten as follows:

$$\sigma_{cr}\sqrt{a} = \sqrt{\frac{2E\gamma}{\pi}} = \text{constant} \tag{2.22}$$

The Griffith formalism, therefore, requires that the quantity $\sigma_{cr}\sqrt{a}$ be a constant. The left-hand side of Eqn. (2.22) represents a crack-driving force, in terms of stress, and the right-hand side represents a material property that governs its resistance to unstable crack growth, or its fracture toughness. From previous consideration of stress concentration, Eqn. (2.12), it may be seen that, as $\rho \to 0$,

$$\sigma_m \approx 2\sigma\sqrt{\frac{a}{\rho}}; \quad \sigma\sqrt{a} \approx \frac{1}{2}\sigma_m\sqrt{\rho} \tag{2.23}$$

Thus, these two concepts are equivalent. In the classical failure context, fracture depends on some critical combination of stress at the crack tip and the tip radius, neither of which are precisely defined (or definable) or accessible to measurement. For experimental accuracy and practical application, it is more appropriate to use the accessible quantities σ and a to determine the fracture toughness of the material. It is to be recognized that the quantities involving $\sigma^2 a$ and $\sigma\sqrt{a}$ represent the crack-driving force, and 2γ, in the Griffith sense, represents the material's resistance to crack growth, or its fracture toughness.

Griffith applied this relationship, Eqn. (2.21), to the study of fracture strengths of glass, and found good agreement with experimental data. The theory did not work well for metals. For example, with $\gamma \approx 1 \text{ J/m}^2$, $E = 210 \text{ GPa}$ and σ_{cr}, fracture is predicted to occur at about yield stress level in mild steels if crack size exceeded about 3 μm. This is contrary to experimental observations that indicated one to two orders of magnitude greater crack tolerance. Thus, Griffith's theory did not find favor in the metals community.

2.4 Modifications to Griffith's Theory

With ship failures during and immediately following World War II, interest in the Griffith theory was revived. Orowan [6] and Irwin [7] both recognized that significant plastic deformation accompanied crack advance in metallic materials, and that the 'plastic work' about the advancing crack contributed to the work required to create new crack surfaces. Orowan suggested that this work might be treated as being equivalent to surface energy (or γ_p), and can be added to the surface energy γ. Thus, the Griffith theory, or fracture criterion, is modified to the following form.

$$\sigma_{cr} = \sqrt{\frac{2E(\gamma + \gamma_p)}{\pi a}} \tag{2.24}$$

This simple addition of γ and γ_p led to conceptual difficulties. Since the nature of the terms are not compatible (the first being a microscopic quantity, and the second, a macroscopic quantity), the addition could not be justified.

It is far more satisfying to simply draw an analogy between the Griffith case for 'brittle' materials and that of more ductile materials. In the later case, it is assumed that if the plastic deformation is sufficiently localized to the crack tip, the crack-driving force may still be characterized in terms of $\mathbf{G}$ from the elasticity analysis. Through the Griffith formalism, a counter part to the crack growth resistance $\mathbf{R}$ can be defined, and the actual value can then be determined by laboratory measurements, and is defined as the fracture toughness $\mathbf{G_c}$. This approach forms the basis for modern day fracture mechanics, and will be considered in detail later.

2.5 Estimation of Crack-Driving Force G from Energy Loss Rate (Irwin and Kies [8, 9])

The crack-driving force G may be estimated from energy considerations. Consider an arbitrarily shaped body containing a crack, with area A, loaded in tension by a force P applied in a direction perpendicular to the crack plane as illustrated in Fig. 2.6. For simplicity, the body is assumed to be pinned at the opposite end. Under load, the stresses in the body will be elastic, except in a small zone near the crack tip (*i.e.*, in the crack-tip plastic zone). If the zone of plastic deformation is small relative to the size of the crack and the dimensions of the body, a linear elastic analysis may be justified as being a good approximation. The stressed body, then, may be characterized by an elastic strain energy function U that depends on the load P and the crack area A (*i.e.*, $U = U(P, A)$), and the elastic constants of the material.

If the crack area enlarges (*i.e.*, the crack grows) by an amount dA, the 'energy' that tends to promote the growth is composed of the work done by the external force P, or $P(d\Delta/dA)$, where Δ is the load-point displacement, and the release in

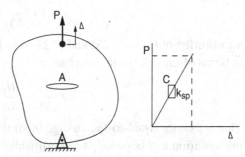

Figure 2.6. A body containing a crack of area A loaded in tension.

strain energy, or $-dU/dA$ (a minus sign is used here because dU/dA represents a decrease in strain energy per unit crack area and is negative). The crack driving force G, by definition, is the sum of these two quantities.

$$G \equiv P\frac{d\Delta}{dA} - \frac{dU}{dA} \qquad (2.25)$$

Because the initial considerations were made under fixed-grip assumptions, where the work by external forces would be zero, the nomenclature *strain energy release rate* is commonly associated with **G**.

Assuming linear elastic behavior, the body can be viewed as a linear spring. The stored elastic strain energy U is given by the applied load (P) and the load-point displacement (Δ), or in terms of the compliance (C) of the body, or the inverse of its stiffness or spring constant; *i.e.*,

$$U = \frac{1}{2}P\Delta = \frac{1}{2}k_{sp}\Delta^2 = \frac{1}{2}P^2C \qquad (2.26)$$

The load-point displacement is equal to the product of P and C; *i.e.*,

$$\Delta = PC \qquad (2.27)$$

The compliance C is a function of crack size, and of the elastic modulus of the material and the dimensions of the body, but, because the latter quantities are constant, C is a function of only A. Thus, $\Delta = \Delta(P, C) = \Delta(P, A)$ and $U = U(P, C) = U(P, A)$.

The work done is given by $Pd\Delta$:

$$Pd\Delta = P\left[\left(\frac{\partial\Delta}{\partial P}\right)_A dP + \left(\frac{\partial\Delta}{\partial A}\right)_P dA\right] = P\left[CdP + P\frac{dC}{dA}dA\right]$$

Thus,

$$P\frac{d\Delta}{dA} = PC\frac{dP}{dA} + P^2\frac{dC}{dA} \qquad (2.28)$$

Similarly,

$$dU = \left(\frac{\partial U}{\partial P}\right)_A dP + \left(\frac{\partial U}{\partial A}\right)_P dA = PCdP + \frac{1}{2}P^2\frac{dC}{dA}dA$$

and

$$\frac{dU}{dA} = PC\frac{dP}{dA} + \frac{1}{2}P^2\frac{dC}{dA} \qquad (2.29)$$

Substitution of Eqns. (2.28) and (2.29) into Eqn. (2.25) gives the crack-driving force in terms of the change in compliance.

$$G = P\frac{d\Delta}{dA} - \frac{dU}{dA} = \frac{1}{2}P^2\frac{dC}{dA} \qquad (2.30)$$

This is exactly equal to the change in strain energy under constant load. Since no precondition was imposed, it is worthwhile to examine the validity of this result for

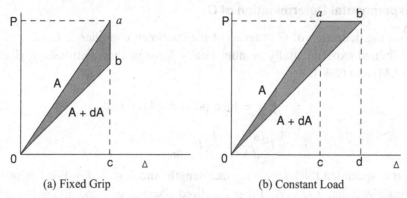

(a) Fixed Grip   (b) Constant Load

Figure 2.7. Load-displacement diagrams showing the source of energy for driving a crack.

the two limiting conditions; *i.e.*, constant load (P = constant) and fixed grip (Δ = constant). Using Eqns (2.25), (2.28), and (2.29), it can be seen that:

$$G_P = \left[P^2 \frac{dC}{dA} + PC \frac{dP}{dA} \right]_P - \left[\frac{1}{2} P^2 \frac{dC}{dA} + PC \frac{dP}{dA} \right]_P = \frac{1}{2} P^2 \frac{dC}{dA} \qquad (2.31)$$

$$G_\Delta = \left[P^2 \frac{dC}{dA} + PC \frac{dP}{dA} \right]_\Delta - \left[\frac{1}{2} P^2 \frac{dC}{dA} + PC \frac{dP}{dA} \right]_\Delta = \frac{1}{2} P^2 \frac{dC}{dA} \qquad (2.32)$$

Thus, the crack-driving force is identical, irrespective of the loading condition.

The source of the energy, however, is different, and may be seen through an analysis of the load-displacement diagrams (Fig. 2.7). Under fixed-grip conditions, the driving force is derived from the release of stored elastic energy with crack extension. It is represented by the shaded area Oab, the difference between the stored elastic energy before and after crack extension (*i.e.*, area Oac and area Obc). For constant load, on the other hand, the energy is provided by the work done by the external force (as represented by the area abcd), minus the increase in the stored elastic energy in the body by $Pd\Delta/2$ (*i.e.*, the difference between areas Obd and Oac); *i.e.*, the shaded area Oab.

It should be noted that G could increase, remain constant, or decrease with crack extension, depending on the type of loading and on the geometry of the crack and the body. For example, it increases for remote tensile loading as depicted on the left of Fig. 2.8, and for wedge-force loading on the right.

Fracture instability occurs when G reaches a critical value:

$$G \to 2\gamma \quad \text{for brittle materials (Griffith crack)}$$

$$G \to G_c \quad \text{for real materials that exhibit some plasticity}$$

Figure 2.8. Examples of crack bodies and loading in which G increases or decreases with crack extension.

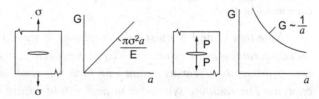

2.6 Experimental Determination of G

Based on the definition of G in terms of the specimen compliance C, G or K may be determined experimentally or numerically through the relationships given by Eqns. (2.33) and (2.34).

$$C = \frac{\Delta}{P}; \Delta = \text{load-point displacement} \tag{2.33}$$

$$G = \frac{1}{2}P^2\frac{dC}{dA} = \frac{1}{2B}P^2\frac{dC}{da} \tag{2.34}$$

where B = specimen thickness; a = crack length; and $Bda = dA$. For this process, it is recognized that $EG = K^2$ for generalized plane stress, and $EG = (1 - v^2)K^2$ for plane strain (to be shown later). It should be noted that the crack-driving force G approaches zero and the crack length a approaches zero. As such, special attention needs to be given to ensure that dC/da also approaches zero in the analysis of experimental or numerical data. The physical processes are illustrated in Fig. 2.9.

The procedure, then, is as follows:

1. Measure the specimen compliance C for various values of crack length a, for a given specimen geometry, from the *LOAD* versus *LOAD-POINT DIS-PLACEMENT* curves. Note that this may be done experimentally or numerically from a finite-element analysis.
2. Construct a C versus a plot and differentiate (graphically, numerically, or by using a suitable curve-fitting routine) to obtain dC/da versus a data.
3. Compute G and K as a function of a through Eqn. (2.34).

Some useful notes:

1. 'Cracks' may be real cracks (such as fatigue cracks) or simulated cracks (*i.e.*, notches). If notches are used, they must be narrow and have well defined, 'rounded' tips.

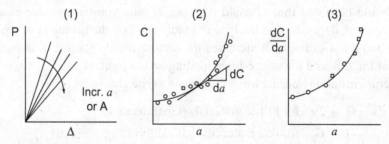

Figure 2.9. Graphical representation of steps in the determination of G or K versus a by the compliance method.

(*Note that G and K must be zero at a = 0; see Eqn. (2.34). As such, the data reduction routine must ensure that dC/da is equal to zero at a = 0. A simple procedure is to combine the C versus a data with their reflection into the second quadrant for analysis. The resulting symmetry in data would ensure that only the even-powered terms would be retained in the polynomial fit, and that dC/da would be zero at a = 0.*)

2. Load-point displacement must be used, since the strain energy for the body is defined as one-half the applied load times this displacement.
3. Instrumentation – load cell, linearly variable differential transformer (LVDT), clip gage, etc.
4. Must have sufficient number of data points to ensure accuracy; particularly for crack length near zero.
5. Accuracy and precision important: must be free from systematic errors; and must minimize variability because of the double differentiation involved in going from Δ versus P, to $C(=\Delta/P)$ versus a, and then to dC/da versus a.
6. Two types of nonlinearities must be recognized and corrected: (i) unavoidable misalignment in the system, and (ii) crack closure. A third type, associated with significant plastic deformation at the 'crack' tip, is not permitted (use of too high a load in calibration).

2.7 Fracture Behavior and Crack Growth Resistance Curve

In the original consideration of fracture, and indeed in the linear elasticity considerations, the crack is assumed to be stationary (*i.e.*, does not grow) up to the point of fracture or instability. If there were a means for monitoring crack extension, say by measuring the opening displacement of the crack faces along the direction of loading, the typical load-displacement curve would be as shown in Fig. 2.10. For a stationary crack in an ideally brittle solid, the load-displacement response would be a straight line (as indicated by the solid line), its slope reflecting the compliance of the cracked body. It should be noted that crack growth in the body would be reflected by a deviation from this linear behavior. This deviation corresponds to an increase in compliance of the body for the longer crack, and is indicated by the dashed line. At a critical load (or at instability), the body simply breaks with a sudden drop-off in load.

The strain energy release G versus crack length a (or stress intensity factor K versus a) space is depicted in Fig. 2.11 for a Griffith crack (*i.e.*, a central through-thickness crack in an infinitely large plate loaded in remote tension in mode I). The change in G with crack length a at a given applied stress σ is indicated by the solid and dashed lines. Because the crack is assumed not to be growing below the critical stress level, the crack growth resistance R is taken to be equal to the driving force G for the initial crack length a_o at each stress level, and is depicted by the vertical line at a_o. At the onset of fracture (or crack growth instability), R is constant and is equal to twice the solid-state surface energy or 2γ. Clearly, in this case, the crack

Figure 2.10. Typical load-displacement curve for an ideally brittle material with a through-thickness crack. Displacement is measured across the crack opening.

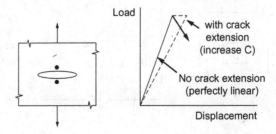

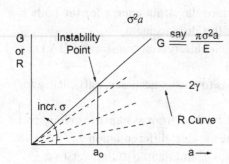

Figure 2.11. Crack growth resistance curve for an ideally brittle material.

growth resistance curve would be independent of crack length, but the critical stress for failure would be a function of the initial crack length as indicated by Eqn. (2.21).

In real materials, however, some deviation from linearity or crack growth would occur with increases in load. They are associated with:

1. apparent crack growth due to crack tip plasticity;
2. adjustment in crack front shape (or crack tunneling) and crack growth associated with increasing load; and
3. crack growth due to environmental influences (stress corrosion cracking) or other time-dependent behavior (creep, etc.).

For fracture over relatively short times (less then tens of seconds) that are associated with the onset of crack growth instability, the time-dependent contributions (item 3) are typically small and may be neglected. The fracture behavior may be considered for the case of a monotonically increasing load.

Recalling the fracture locus in terms of stress (or load) versus crack length (σ versus a) discussed in Chapter 1 (Fig. 1.7), the fracture behavior may be considered in relation to the three regions (A, B, and C) of response (Fig. 2.12). Region A is considered to extend from stress levels equal to the tensile yield strength (σ_{YS}) to the ultimate (or 'notch') tensile strength (σ_{UTS} σ_{NTS}); region B, for stresses from about σ_{YS} to $0.8\sigma_{YS}$; and region C, for stresses below $0.8\sigma_{YS}$.

REGION A: Failure occurs by general yielding and is associated with large extension as if no crack is present. The load-displacement response is schematically indicated in Fig. 2.13 along with a typical failed specimen. Yielding extends across the entire uncracked section, and the displacement is principally associated with plastic extension. Fracture is characterized by considerable contractions

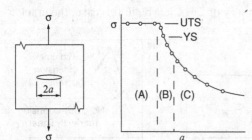

Figure 2.12. Failure locus in terms of stress versus crack length separated into three regions (A, B, and C) of response.

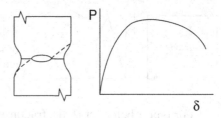

Figure 2.13. A schematic illustration of the load-displacement curve and a typical example of a specimen fractured in Region A.

(or 'necking') in both the width and thickness directions. Because of the presence of the crack, failure still tends to proceed outward either along the original crack direction or by shearing along an oblique plane (see Fig. 2.13). Because of the large-scale plastic deformation associated with fracture, this region is not of interest to LEFM and will not be considered further.

REGION B: This is the transition region between what is commonly (although imprecisely) referred to as 'ductile' and 'brittle' fracture. In a continuum sense, it is a region between fracture in the presence of large-scale plastic deformation and one in which plastic deformation is limited to a very small region at the crack tip. Crack growth in this region occurs with the uncracked section near or at yielding (*i.e.*, with $0.8\sigma_{YS} < \sigma < \sigma_{YS}$). The load-displacement response is schematically indicated in Fig. 2.14 along with a typical failed specimen. The load-displacement curves would reflect contributions of plastic deformation as well as crack growth. Since the plastically deformed zone represents an appreciable fraction of the uncracked section, and is large in relation to the crack size, this region is also not of interest to LEFM. From a practical viewpoint, however, this region is of considerable importance for low-strength–high-toughness materials, and is treated by elastic-plastic fracture mechanics (EPFM).

REGION C: Fracture in this region is commonly considered to be 'brittle' (in the continuum sense). The zone of plastic deformation at the crack tip is small relative to the size of the crack and the uncracked (or net) section. The stress at fracture is often well below the tensile yield strength. The load-displacement response exhibits two typical types of behavior, depending on the material thickness, that are illustrated in Fig. 2.15. Type 1 behavior corresponds to thicker materials and reflects the limited plastic deformation (or a more "brittle" response) that accompanies fracture. Type 2, for thinner materials, on the other hand, reflects the evolution of increased resistance (or a more "ductile" response) to unstable crack growth with crack prolongation and the associated crack-tip plastic deformation under an increasing applied load (see Fig. 2.16). Description of fracture behavior in this region is the principal domain of LEFM.

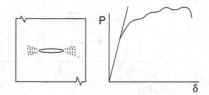

Figure 2.14. A schematic illustration of the load-displacement curve and a typical example of a specimen fractured in Region B.

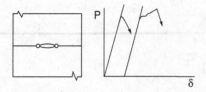

Figure 2.15. A schematic illustration of the two types of load-displacement curves for specimens of different thickness fractured in Region C.

For type 1 behavior (left), fracture is abrupt, nonlinearity is associated with the development of the crack-tip plastic zone. For type 2 behavior (right), on the other hand, each point along the load-displacement curve would correspond to a different effective crack length, which corresponds to the actual physical crack length plus a 'correction' for the zone of crack-tip plastic deformation (see Chapter 4). In practice, if one unloads from any point on the load-displacement curve, the unloading slope would reflect the unloading compliance, or the physical crack length, at that point, and the intercept would represent the contribution of the crack-tip plastic zone. In other words, the line that joins that point with the origin of the load-displacement curve would reflect the effective crack length of the point. Again, based on the effective crack length and the applied load (or stress), the crack-driving force G or K could be calculated for that point. Since the crack would be in stable equilibrium, in the absence of time-dependent effects (*i.e.*, with G in balance with the crack growth resistance R) at that point, R is equal to G (or $K_R = K$). By successive calculations, a crack growth resistance curve (or R curve) can be constructed in the G versus a, or R versus a, space, Fig. 2.16b. The crack growth instability point is then the point of tangency between the G (for the critical stress) and R, or K and K_R, curves. The value of R, or K_R, at instability is defined as the fracture toughness G_c, or K_c. (Note that, in fracture toughness testing, both the load and crack length at the onset of instability must be measured.) Available evidence (see ASTM STP 527 [10]) indicates that R is only a function of crack extension (Δa) rather than the actual crack length; in other words, R depends on the evolution of resistance with crack extension. It may be seen readily from Fig. 2.17 that the fracture toughness G_c, or K_c. is expected to depend on crack length. For this reason, the use of R curves in design is preferred.

In principle then, a fracture toughness parameter has been defined in terms of linear elastic analysis of a cracked body involving the strain energy release rate G, or the stress intensity factor K. For thick sections, the fracture toughness is defined as G_{Ic}, and for thinner sections, as G_c or R (referred only to mode I loading here). This value is to be measured in the laboratory and applied to design. The validity of

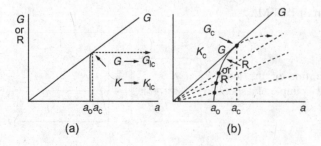

Figure 2.16. Crack growth resistance curves associated with Types 1 and 2 load-displacement response in Region C (Fig. 2.15): (a) for Type 1 response associated with thicker materials; (b) Type 2 response for thinner materials.

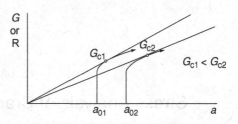

Figure 2.17. Schematic illustration showing the expected dependence of G_c on crack length a.

this measurement and its utilization depends on the ability to satisfy the assumption of limited plasticity that is inherent in the use of linear elasticity analysis. This issue will be taken up after a more formalized consideration of the stress analysis of a cracked body in Chapter 3.

REFERENCES

[1] Griffith, A. A., "The Phenomenon of Rupture and Flow in Solids," Phil. Trans. Royal Soc. of London, A221 (1921), 163–197.

[2] Griffith, A. A., "The Theory of Rupture," Proc. 1st Int. Congress Applied Mech. (1924), 55–63. Biezeno and Burgers, eds., Waltman (1925).

[3] Tresca, H., "On the "flow of solids" with practical application of forgings, etc.," Proc. Inst. Mech. Eng., 18 (1867), 114–150.

[4] Von Mises, R., "Mechanik der plastischen Formänderung von Kristallen," ZAMM-Zeitschrift für Angewandte Mathematik und Mechanik, 8, 3 (1928), 161–185.

[5] Inglis, C. E., "Stresses in a Plate due to the Presence of Cracks and Sharp Corners," Trans. Inst. Naval Architects, 55 (1913), 219–241.

[6] Orowan, E., "Energy Criterion of Fracture," Welding Journal, 34 (1955), 1575–1605.

[7] Irwin, G. R., "Fracture Dynamics," in Fracturing of Metals, ASM publication (1948), 147–166.

[8] Irwin, G. R., and Kies, J. A., "Fracturing and Fracture Dynamics," Welding Journal Research Supplement (1952).

[9] Irwin, G. R., and Kies, J. A., "Critical Energy Rate Analysis of Fracture Strength of Large Welded Structures," The Welding Journal Research Supplement (1954).

[10] ASTM STP 527, Fracture Toughness Evaluation by R-Curve Method, American Society for Testing and Materials, Philadelphia, PA (1973).

3 Stress Analysis of Cracks

Traditionally, design engineers prefer to work with stresses rather than energy, or energy release rates. As such, a shift in emphasis from energy to the stress analysis approach was made in the late 1950s, starting with Irwin's paper [1], published in the Journal of Applied Mechanics of ASME. In this paper, Irwin demonstrated the equivalence between the stress analysis and strain energy release rate approaches. This seminal work was followed by a wealth of papers over the succeeding decades that provided linear elasticity-based, stress intensity factor solutions for cracks and loadings of nearly every conceivable shape and form. Analytical (or closed-form) solutions were obtained for the simpler geometries and configurations, and numerical solutions were provided, or could be readily obtained with modern finite-element analysis codes, for the more complex cases. Most of the solutions are available in handbooks (*e.g.*, Sih [2]; Tada *et al.* [3]; Broek [4]). Others can be obtained by superposition, or through the use of computational techniques.

Most of the crack problems that have been solved are based on *two-dimensional*, linear elasticity (*i.e.*, the infinitesimal or small strain theory for elasticity). Some *three-dimensional* problems have also been solved; however, they are limited principally to axisymmetric cases. Complex variable techniques have served well in the solution of these problems. To gain a better appreciation of the problems of fracture and crack growth, it is important to understand the basic assumptions and ramifications that underlie the stress analysis of cracks.

3.1 Two-Dimensional Theory of Elasticity

To provide this basic appreciation, a brief review of two-dimensional theory of elasticity is given below, followed by a summary of the basic formulation of the crack problem. More complete treatments of the theory of elasticity may be found in standard textbooks and other treatises (*e.g.*, Mushkilishevili [5]; Sokolnikoff [6]; Timoshenko [7]).

3.1.1 Stresses

Stress, in its simplest term, is defined as the force per unit area over a surface as the surface area is allowed to be reduced, in the limit, to zero. Mathematically, stress is expressed as follows:

$$\sigma = \lim_{\Delta A \to 0} \frac{\Delta F}{\Delta A} \tag{3.1}$$

where ΔF is the force over an increment of area ΔA.

In general, the stresses at a point are resolved into nine components. In Cartesian coordinates, these include the three normal components σ_{xx}, σ_{yy}, and σ_{zz}, and the shearing components τ_{xy}, τ_{xz}, τ_{yz}, τ_{yx}, τ_{zx}, and τ_{zy}, and may be given in matrix form as follows:

$$\begin{pmatrix} \sigma_{xx} & \tau_{xy} & \tau_{xz} \\ \tau_{yx} & \sigma_{yy} & \tau_{yz} \\ \tau_{zx} & \tau_{zy} & \sigma_{zz} \end{pmatrix} \tag{3.2}$$

The first letter in the subscript designates the plane on which the stress is acting, and the second designates the direction of the stress.

For two-dimensional problems, two special cases are considered; namely, *plane stress* and *plane strain*. For the case of plane stress, only the in-plane (*e.g.*, the *xy*-plane) components of the stresses are nonzero; and for plane strain, only the in-plane components of strains are nonzero. In reality, however, only the average values of the *z*-component stresses are zero in the "plane stress" cases. As such, this class of problems is designated by the term *generalized plane stress*. The conditions for each case will be discussed later. It is to be recognized that, in actual crack problems, these limiting conditions are never achieved. References to plane stress and plane strain, therefore, always connote approximations to these well-defined conditions.

3.1.2 Equilibrium

There are nine components of (unknown) stresses at any point in a stressed body, and they generally vary from point to point within the body. These stresses must be in equilibrium with each other and with other body forces (such as gravitational and inertial forces). For elastostatic problems, the body forces are typically assumed to be zero, and are not considered further. For simplicity, therefore, the equilibrium of an element (dx, dy, 1) under plane stress ($\sigma_{zz} = \tau_{zx} = \tau_{xz} = \tau_{zy} = \tau_{yz} = 0$) is considered, as depicted in Fig. 3.1.

The changes in stress with position are represented by the Taylor series expansions shown, with the higher-order terms in the series neglected.

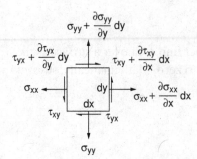

Figure 3.1. Equilibrium of stresses at a point under the state of plane stress.

Neglecting the body forces, equilibrium conditions require that the summation of moment and forces to be zero; *i.e.*:

$$\sum M_A = 0 = \left[\tau_{xy} + \tau_{xy} + \frac{\partial \tau_{xy}}{\partial x} dx\right] dy \frac{dx}{2} - \left[\tau_{yx} + \tau_{yx} + \frac{\partial \tau_{yx}}{\partial y} dy\right] dx \frac{dy}{2}$$

$$\sum F_x = 0 = \left[\sigma_{xx} + \frac{\partial \sigma_{xx}}{\partial x} dx - \sigma_{xx}\right] dy + \left[\tau_{yx} + \frac{\partial \tau_{yx}}{\partial y} dy - \tau_{yx}\right] dx \qquad (3.3)$$

$$\sum F_y = 0 = \left[\sigma_{yy} + \frac{\partial \sigma_{yy}}{\partial y} dy - \sigma_{yy}\right] dx + \left[\tau_{xy} + \frac{\partial \tau_{xy}}{\partial x} dx - \tau_{xy}\right] dy$$

The first of these equilibrium conditions leads to the fact that the shearing stresses must be equal; *i.e.*, $\tau_{xy} = \tau_{yx}$. The next two lead to the following two equilibrium equations:

$$\frac{\partial \sigma_{xx}}{\partial x} + \frac{\partial \tau_{yx}}{\partial y} = 0$$

$$\frac{\partial \tau_{xy}}{\partial x} + \frac{\partial \sigma_{yy}}{\partial y} = 0 \qquad (3.4)$$

3.1.3 Stress-Strain and Strain-Displacement Relations

The strains at a point are resolved into nine components. In Cartesian coordinates, these include the three normal components, ε_{xx}, ε_{yy}, and ε_{zz}, and the shearing components γ_{xy}, γ_{xz}, γ_{yz}, γ_{yx}, γ_{zx}, and γ_{zy}, and may be given in matrix form as follows:

$$\begin{pmatrix} \varepsilon_{xx} & \gamma_{xy} & \gamma_{xz} \\ \gamma_{yx} & \varepsilon_{yy} & \gamma_{yz} \\ \gamma_{zx} & \gamma_{zy} & \varepsilon_{zz} \end{pmatrix} \qquad (3.5)$$

Only six components of strain, however, are required because $\gamma_{xy} = \gamma_{yx}$, $\gamma_{yz} = \gamma_{zy}$, and $\gamma_{zx} = \gamma_{xz}$. The six components of strains are related to the six components of stresses through Hooke's law; namely,

$$\varepsilon_{xx} = \frac{1}{E}[\sigma_{xx} - v\sigma_{yy} - v\sigma_{zz}]$$

$$\varepsilon_{yy} = \frac{1}{E}[\sigma_{yy} - v\sigma_{zz} - v\sigma_{xx}]$$

$$\varepsilon_{zz} = \frac{1}{E}[\sigma_{zz} - v\sigma_{xx} - v\sigma_{yy}]$$

$$\gamma_{xy} = \frac{2(1+v)}{E}\tau_{xy} = \frac{1}{\mu}\tau_{xy} \qquad\qquad (3.6)$$

$$\gamma_{yz} = \frac{2(1+v)}{E}\tau_{yz} = \frac{1}{\mu}\tau_{yz}$$

$$\gamma_{zx} = \frac{2(1+v)}{E}\tau_{zx} = \frac{1}{\mu}\tau_{zx}$$

Here, E and μ are the elastic (Young's) and shearing modulus, respectively, where $E = 2(1 + v)\mu$; and v is the Poisson ratio. In terms of two-dimensional problems, there are now six unknowns (three components of stresses and three components of strains) related through five independent equations; i.e., the two equations of equilibrium and three stress-strain relationships (or Hooke's law). For three-dimensional problems, on the other hand, the number of unknowns is twelve; these unknowns are related at this point through three equations of equilibrium and six stress-strain relationships.

To proceed further, one can consider the displacements $u = u(x, y)$ and $v = v(x, y)$, which are functions only of the in-plane coordinates x and y in two-dimensional problems. It can be readily shown that the displacements are related to the strains through the following relationships:

$$\varepsilon_{xx} = \frac{\partial u}{\partial x}$$

$$\varepsilon_{yy} = \frac{\partial v}{\partial y} \qquad\qquad (3.7)$$

$$\gamma_{xy} = \frac{\partial u}{\partial y} + \frac{\partial v}{\partial x}$$

Note that the out-of-plane or z-component of displacement, $w = w(x, y)$, depends also only on x and y here, and does not contribute to strain.

There are now eight equations with eight unknowns (stresses, strains, and displacements) that are interrelated. The three components of strains are related to the two displacement components and, therefore, cannot be taken arbitrarily. The solution of two-dimensional elasticity problems, therefore, requires an additional independent equation.

3.1.4 Compatibility Relationship

Solution of elasticity problems is constrained by the requirement that the strains must be continuous, which means that the deformation or strains within the body must be 'compatible' with each other. Continuity, or compatibility, in strains, in turn, requires the strains to have continuous derivatives. By differentiating ε_{xx} twice with respect to y, ε_{yy} twice with respect to x, and γ_{xy} with respect to x and y, the following relationships are obtained:

$$\frac{\partial^2 \varepsilon_{xx}}{\partial y^2} = \frac{\partial^3 u}{\partial x \partial y^2}; \quad \frac{\partial^2 \varepsilon_{yy}}{\partial x^2} = \frac{\partial^3 v}{\partial x^2 \partial y}; \quad \frac{\partial^2 \gamma_{xy}}{\partial x \partial y} = \frac{\partial^3 u}{\partial x \partial y^2} + \frac{\partial^3 v}{\partial x^2 \partial y} \qquad (3.8)$$

An examination of Eqn. (3.8) shows that the individual relations may be combined into a single relationship, the *compatibility relationship*, as follows:

$$\frac{\partial^2 \varepsilon_{xx}}{\partial y^2} + \frac{\partial^2 \varepsilon_{yy}}{\partial x^2} = \frac{\partial^2 \gamma_{xy}}{\partial x \partial y} \qquad (3.9)$$

This relationship guarantees continuity in displacements and uniqueness of the solution.

3.2 Airy's Stress Function

Thus, the solution of two-dimensional elastostatic problems reduces to the integration of the equations of equilibrium together with the compatibility equation, and to satisfy the boundary conditions. The usual method of solution is to introduce a new function (commonly known as Airy's stress function), and is outlined in the next subsections.

3.2.1 Basic Formulation

Airy's stress function $\Phi(x, y)$ is related to the stresses as follows:

$$\sigma_{xx} = \frac{\partial^2 \Phi}{\partial y^2}$$

$$\sigma_{yy} = \frac{\partial^2 \Phi}{\partial x^2} \qquad (3.10)$$

$$\tau_{xy} = -\frac{\partial^2 \Phi}{\partial x \partial y}$$

By substituting these relationships into the equilibrium equations (3.4) and performing the indicated differentiation, it can be readily shown that the equilibrium conditions are automatically satisfied by this function (see Eqn. (3.11)).

$$\frac{\partial \sigma_{xx}}{\partial x} + \frac{\partial \sigma_{yx}}{\partial y} = \frac{\partial^3 \Phi}{\partial x \partial y^2} - \frac{\partial^3 \Phi}{\partial x \partial y^2} = 0$$

$$\frac{\partial \tau_{xy}}{\partial x} + \frac{\partial \sigma_{yy}}{\partial y} = -\frac{\partial^3 \Phi}{\partial x^2 \partial y} + \frac{\partial^3 \Phi}{\partial x^2 \partial y} = 0 \qquad (3.11)$$

The compatibility equation may now be written in terms of Airy's stress function through the use of the stress-strain relationships as follows:

$$\frac{\partial^2 \varepsilon_{xx}}{\partial y^2} + \frac{\partial^2 \varepsilon_{yy}}{\partial x^2} = \frac{\partial^2 \gamma_{xy}}{\partial x \partial y}$$

$$\frac{1}{E}\left[\frac{\partial^2 \sigma_{xx}}{\partial y^2} - v\frac{\partial^2 \sigma_{yy}}{\partial y^2} + \frac{\partial^2 \sigma_{yy}}{\partial x^2} - v\frac{\partial^2 \sigma_{xx}}{\partial x^2}\right] = \frac{2(1+v)}{E}\frac{\partial^2 \tau_{xy}}{\partial x \partial y}$$

$$\frac{\partial^4 \Phi}{\partial y^4} - v\frac{\partial^4 \Phi}{\partial x^2 \partial y^2} + \frac{\partial^4 \Phi}{\partial x^4} - v\frac{\partial^4 \Phi}{\partial x^2 \partial y^2} = -2(1+v)\frac{\partial^4 \Phi}{\partial x^2 \partial y^2}$$

Therefore,

$$\frac{\partial^4 \Phi}{\partial x^4} + 2\frac{\partial^4 \Phi}{\partial x^2 \partial y^2} + \frac{\partial^4 \Phi}{\partial y^4} = 0 \tag{3.12}$$

Equation (3.12) is the governing partial differential equation for two-dimensional elasticity. Any function that satisfies this fourth-order partial differential equation will satisfy all of the eight equations of elasticity; *namely*, the equilibrium equations, Hooke's law, and the strain-displacement relations.

The governing differential equation may be rewritten in more compact form by considering the differential operator ∇^2, where:

$$\nabla^2 = \frac{\partial^2}{\partial x^2} + \frac{\partial^2}{\partial y^2} \tag{3.13}$$

Operating on the function $\Phi(x, y)$ twice yields:

$$\nabla^2 \Phi = \frac{\partial^2 \Phi}{\partial x^2} + \frac{\partial^2 \Phi}{\partial y^2}$$

$$\nabla^4 \Phi = \nabla^2(\nabla^2 \Phi) = \frac{\partial^4 \Phi}{\partial x^4} + 2\frac{\partial^4 \Phi}{\partial x^2 \partial y^2} + \frac{\partial^4 \Phi}{\partial y^4}$$

The governing differential equation, therefore, may be written in the following form:

$$\nabla^4 \Phi(x, y) = 0 \quad \text{(biharmonic equation)} \tag{3.14}$$

The solution of plane (two-dimensional) elasticity problem now resides in the determination of an Airy stress function $\Phi(x, y)$ that satisfies the governing fourth-order partial differential equation and the appropriate boundary conditions. Note that:

$$\nabla^2(\nabla^2 \Phi) = \nabla^2\left(\frac{\partial^2 \Phi}{\partial x^2} + \frac{\partial^2 \Phi}{\partial y^2}\right) = \nabla^2(\sigma_{yy} + \sigma_{xx}) = 0$$

The sum of the stresses $(\sigma_{xx} + \sigma_{yy})$, therefore, must be harmonic.

The function $\Phi(x, y)$ may be chosen to be a linear combination of functions of the form:

$$\Phi(x, y) = \Psi_1 + x\Psi_2 + y\Psi_3 + (x^2 + y^2)\Psi_4 \tag{3.15}$$

The function $\Phi(x, y)$ would satisfy the biharmonic Eqn. (3.14) if each of the functions $\Psi_i(x, y)$ is harmonic; *i.e.*, they satisfy the equation:

$$\nabla^2 \Psi_i(x, y) = 0 \tag{3.16}$$

This, in essence, is the application of the well-known principle of superposition.

3.2.2 Method of Solution Using Functions of Complex Variables

The use of complex variables and complex functions provides a powerful technique for solving problems in two-dimensional elasticity. The reader is encouraged to consult the many textbooks and treatises on these subjects. An abbreviated treatment is given here as a lead-in for the consideration of stresses and strains near the tip of a stationary crack.

Complex Numbers

A complex number $a + ib$ is composed of a real part a and an imaginary part b, with the imaginary part defined through the use of $i = \sqrt{-1}$. The addition, subtraction, multiplication, division, and taking roots follow conventional rules in which the real and imaginary parts are kept separate. For example:

Addition: $(a + ib) + (c + id) = (a + c) + i(b + d)$

Multiplication: $(a + ib)(c + id) = (ac - bd) + i(ad + bc)$

Division: $\dfrac{(a + ib)}{(c + id)} = \dfrac{(a + ib)}{(c + id)} \dfrac{(c - id)}{(c - id)} = \dfrac{(ac + bd) - i(ad - bc)}{(c^2 + d^2)}$

Roots: $\sqrt{(a + ib)} = p + iq$ (a real and an imaginary part)

$(a + ib) = (p + iq)^2 = (p^2 - q^2) + 2ipq$

where $a = (p^2 - q^2)$ and $b = 2pq$ are to be solved for p and q.

Complex Variables and Functions

A complex variable $z = x + iy$ can be defined to represent a point in two-dimensional space, where the x-axis is taken to be real, and the y-axis is imaginary. A function of a complex variable $f(z)$ can then be defined, where:

$$f(z) = f(x + iy) = \Re e\, f(z) + i \Im m\, f(z) \tag{3.17}$$

where $\Re e\, f(z)$ and $\Im m\, f(z)$ are the real and imaginary parts of the function $f(z)$. The derivative of $f(z)$ may be similarly defined:

$$f'(z) = \frac{df(z)}{dz} = \Re e\, f'(z) + i \Im m\, f'(z) \tag{3.18}$$

For example, if $f(z) = z^2$, then

$$f(z) = z^2 = (x + iy)^2 = (x^2 - y^2) + 2ixy$$

$$f'(z) = \frac{df(z)}{dz} = 2z = 2(x + iy) = 2x + 2iy$$

Cauchy-Riemann Conditions and Analytic Functions

But $dz = dx + i\,dy$ can be obtained in different ways. By taking $dz = dx$ (*i.e.*, by letting $dy = 0$) and applying the chain rule of differentiation, then:

$$f'(z) = \frac{\partial}{\partial x}\Re e\, f(z) + i\frac{\partial}{\partial x}\Im m\, f(z) \equiv \Re e\, f'(z) + i\Im m\, f'(z)$$

Similarly, by taking $dz = i\,dy$ (with $dx = 0$), then:

$$f'(z) = \frac{\partial}{i\partial y}\Re e\, f(z) + i\frac{\partial}{i\partial y}\Im m\, f(z) = \frac{\partial}{\partial y}\Im m\, f(z) - i\frac{\partial}{\partial y}\Re e\, f(z)$$

$$\equiv \Re e\, f'(z) + i\,\Im m\, f'(z)$$

Based on the foregoing example, the derivatives of $f(z)$ by the two procedures are $2x + 2iy$ and $2iy + 2x$, respectively.

For the derivative to exist, these two derivatives must be equal, therefore:

$$\frac{\partial}{\partial x}\Re e\, f(z) = \frac{\partial}{\partial y}\Im m\, f(z)$$

$$\frac{\partial}{\partial x}\Im m\, f(z) = -\frac{\partial}{\partial y}\Re e\, f(z)$$

(3.19)

These are the Cauchy-Riemann conditions. Functions that satisfy the Cauchy-Riemann conditions are called analytic functions. The Cauchy-Riemann conditions are satisfied by any analytic function and, hence, any of its successive derivatives. This property of analytic functions makes them useful in the solution of problems in two-dimensional elasticity.

Considering Eqn. (3.14) and the operation on the real and imaginary parts of an analytic function $f(z)$, one can see that by using the definition of derivatives and the Cauchy-Riemann conditions:

$$\nabla^2\Re e\, f(z) = \left(\frac{\partial^2}{\partial x^2} + \frac{\partial^2}{\partial y^2}\right)\Re e\, f(z) = \frac{\partial}{\partial x}\left(\frac{\partial}{\partial x}\Re e\, f(z)\right) + \frac{\partial}{\partial y}\left(\frac{\partial}{\partial y}\Re e\, f(z)\right)$$

$$= \frac{\partial}{\partial x}\left(\Re e\, f'(z)\right) + \frac{\partial}{\partial y}\left(-\Im m\, f'(z)\right) = \Re e\, f''(z) - \frac{\partial}{\partial x}\left(\Re e\, f'(z)\right)$$

$$= \Re e\, f''(z) - \Re e\, f''(z) = 0$$

$$\nabla^2\Im m\, f(z) = \left(\frac{\partial^2}{\partial x^2} + \frac{\partial^2}{\partial y^2}\right)\Im m\, f(z) = \frac{\partial}{\partial x}\left(\frac{\partial}{\partial x}\Im m\, f(z)\right) + \frac{\partial}{\partial y}\left(\frac{\partial}{\partial y}\Im m\, f(z)\right)$$

$$= \frac{\partial}{\partial x}\left(\Im m\, f'(z)\right) + \frac{\partial}{\partial y}\left(\Re e\, f'(z)\right) = \Im m\, f''(z) - \frac{\partial}{\partial x}\left(\Im m\, f'(z)\right)$$

$$= \Im m\, f''(z) - \Im m\, f''(z) = 0$$

or,

$$\nabla^2\Re e\, f(z) = 0 \quad \text{and} \quad \nabla^2\Im m\, f(z) = 0$$

(3.20)

In other words, the real and imaginary parts of analytic functions are harmonic and would satisfy the biharmonic equation (see Eqn. (3.14)). The task then becomes one of identifying the appropriate analytic functions that can satisfy the boundary conditions of the problem. The methodology is applied to the solution of crack problems.

3.3 Westergaard Stress Function Approach [8]

The Westergaard stress function approach is used most frequently in the solution of crack problems. Although the method has been criticized by Sih [2], it is nevertheless useful in demonstrating the essential features of the problem and the solution methodology, recognizing that the boundary conditions in certain specific cases, involving external loads at infinity, are not generally satisfied. Following Westergaard, the Airy stress function $\Phi(z)$ is chosen, where:

$$\Phi(z) = \Re e\,\overline{\overline{Z}}(z) + y\Im m\,\overline{Z}(z) \tag{3.21}$$

The function $\overline{\overline{Z}}(z)$ is analytic and therefore satisfies the relationship $\nabla^2 \overline{\overline{Z}}(z) = 0$. The derivatives of the function $\overline{\overline{Z}}(z)$ are defined as follows:

$$\overline{Z}(z) = \frac{d\overline{\overline{Z}}(z)}{dz}$$

$$Z(z) = \frac{d\overline{Z}(z)}{dz} \tag{3.22}$$

$$Z'(z) = \frac{dZ(z)}{dz}$$

Because the derivatives of analytic functions are also analytic and are harmonic, the chosen function $\Phi(z)$ satisfies the biharmonic equation $\nabla^4 \Phi = 0$. Note also that Eqn. (3.21) is a special case of Eqn. (3.15), in which only the first and third functions are retained, namely:

$$\Psi_1(z) = \Re e\,\overline{\overline{Z}}(z), \quad \Psi_3(z) = \Im m\,\overline{Z}(z), \quad \text{and} \quad \Psi_2(z) = \Psi_4(z) = 0$$

3.3.1 Stresses

Based on the definition of stresses in terms of the Airy stress function, given in Eqn. (3.10), one obtains from the Westergaard function:

$$\sigma_{xx} = \frac{\partial^2 \Phi}{\partial y^2} = \frac{\partial}{\partial y}\left[\frac{\partial}{\partial y}\left(\Re e\,\overline{\overline{Z}}(z) + y\Im m\,\overline{Z}(z)\right)\right]$$

$$= \frac{\partial}{\partial y}\left[-\Im m\,\overline{Z}(z) + \Im m\,\overline{Z}(z) + y\frac{\partial}{\partial y}\Im m\,\overline{Z}(z)\right]$$

$$= \frac{\partial}{\partial y}\left[y\Re e\,Z(z)\right] = \Re e\,Z(z) - y\Im m\,Z'(z) \tag{3.23a}$$

$$\sigma_{yy} = \frac{\partial^2 \Phi}{\partial x^2} = \frac{\partial}{\partial x}\left[\frac{\partial}{\partial x}\left(\Re e\,\overline{\overline{Z}}(z) + y\Im m\,\overline{Z}(z)\right)\right]$$

$$= \frac{\partial}{\partial x}\left[\Re e\,\overline{Z}(z) + y\frac{\partial}{\partial x}\Im m\,\overline{Z}(z)\right]$$

$$= \frac{\partial}{\partial x}\left[\Re e\,\overline{Z}(z) + y\Im m\,Z(z)\right] = \Re e\,Z(z) + y\Im m\,Z'(z) \qquad (3.23b)$$

$$\tau_{xy} = -\frac{\partial^2 \Phi}{\partial x\partial y} = -\frac{\partial}{\partial x}\left[\frac{\partial}{\partial y}\left(\Re e\,\overline{\overline{Z}}(z) + y\Im m\,\overline{Z}(z)\right)\right]$$

$$= -\frac{\partial}{\partial x}\left[-\Im m\,\overline{Z}(z) + \Im m\,\overline{Z}(z) + y\frac{\partial}{\partial y}\Im m\,\overline{Z}(z)\right] = -y\Re e\,Z'(z) \qquad (3.23c)$$

In summary, the stresses are as follows:

$$\sigma_{xx} = \Re e\,Z(z) - y\Im m\,Z'(z)$$

$$\sigma_{yy} = \Re e\,Z(z) + y\Im m\,Z'(z)$$

$$\sigma_{zz} = \begin{cases} 0; & \text{for plane stress} \\ v(\sigma_{xx} + \sigma_{yy}); & \text{for plane strain} \end{cases} \qquad (3.24)$$

$$\tau_{xy} = -y\Re e\,Z'(z)$$

3.3.2 Displacement (Generalized Plane Stress)

Based on the definition of strains in terms of displacements (see Eqn. (3.7)), and by using Hooke's law (see Eqn. (3.6)), the displacement field in a cracked body can be readily obtained through direct integration. For simplicity, the case of generalized plane stress, involving the in-plane displacements $u(x, y)$ and $v(x, y)$, is considered here. From Hooke's law and the stresses given in Eqn. (3.24), the displacements are obtained as follows:

$$E\varepsilon_{xx} = E\frac{\partial u}{\partial x} = \sigma_{xx} - v\sigma_{yy} = [\Re e\,Z(z) - y\Im m\,Z'(z)] - v[\Re e\,Z(z) + y\Im m\,Z'(z)]$$

$$= (1 - v)\Re e\,Z(z) - (1 + v)y\Im m\,Z'(z)$$

$$Eu = (1 - v)\Re e\,\overline{Z}(z) - (1 + v)y\Im m\,Z(z) + f_1(y) + constant \qquad (3.25a)$$

$$E\varepsilon_{yy} = E\frac{\partial v}{\partial y} = [\Re e\,Z(z) + y\Im m\,Z'(z)] - v[\Re e\,Z(z) - y\Im m\,Z'(z)]$$

$$= (1 - v)\Re e\,Z(z) + (1 + v)y\Im m\,Z'(z) \qquad (3.25b)$$

$$Ev = (1 - v)\Im m\,\overline{Z}(z) + (1 + v)y(-\Re e\,Z(z)) + (1 + v)\int \Re e\,Z(z)dy$$

$$= (1 - v)\Im m\,\overline{Z}(z) + (1 + v)\Im m\,\overline{Z}(z) - (1 + v)y\Re e\,Z(z) + f_2(x) + constant$$

$$= 2\Im m\,\overline{Z}(z) - (1 + v)y\Re e\,Z(z) + f_2(x) + constant$$

The constants in Eqns. (3.25a) and (3.25b) represent rigid-body translations in the x and y directions, respectively, and need not be considered in the calculation of stresses and strains. The other "constants" of integration $f_1(y)$ and $f_2(x)$ are constrained through the shearing strain γ_{xy}. Using the results given in Eqn. (3.25), it follows that

$$E\gamma_{xy} = E\left(\frac{\partial u}{\partial y} + \frac{\partial v}{\partial x}\right) = -2(1+v)y\Re e\,Z'(z) + \frac{\partial f_1(y)}{\partial y} + \frac{\partial f_2(x)}{\partial x} \qquad (3.26)$$

In summary, the in-plane displacements for generalized plane stress are as follows:

$$\begin{aligned}
Eu &= (1-v)\Re e\,\overline{Z}(z) - (1+v)y\Im m\,Z(z) + f_1(y) \\
Ev &= 2\Im m\,\overline{Z}(z) - (1+v)y\Re e\,Z(z) + f_2(x)
\end{aligned} \qquad (3.27)$$

The functions $f_1(y)$ and $f_2(x)$ depend on the boundary conditions and cannot be chosen arbitrarily. It will be shown later, for example, that for a cracked plate under remote biaxial tension, the boundary condition on shearing stresses ($\tau_{xy} = 0$ at infinity) requires that the sum of the last two terms be zero. As such, the derivatives must be equal to a constant of opposite sign, which leads to $f_1(y) = Ay$ and $f_2(x) = -Ax$. Here, the contribution of $f_1(y)$ and $f_2(x)$ to the u and v displacements represents rigid body rotation and is not considered further in the stress analysis of cracked bodies. It should be noted that the constant A cannot be arbitrarily neglected.

3.3.3 Stresses at a Crack Tip and Definition of Stress Intensity Factor

By using the Westergaard approach and the Airy stress function, the stresses near the tip of a crack may be considered (Fig. 3.2). A set of in-plane Cartesian coordinates x and y, or polar coordinates r and θ, is chosen, with the origin at the crack tip. The boundary conditions are as follows: (i) stresses at the crack tip are very large; and (ii) the crack surfaces are stress free.

Along the $y = 0$ plane, the normal stresses σ_{xx} and σ_{yy} would be given by Eqn. (3.24) as

$$\sigma_{xx} = \sigma_{yy} = \Re e\,Z(z)$$

The physical requirement that strain energy in the elastic body be finite (or bounded) suggests that the order of singularity of stresses at the crack tip can be represented at most by $z^{-1/2}$. (The basic reasoning is that, with stress and strain proportional to $z^{-1/2}$, the strain energy density would be proportional to r^{-1}. The strain energy in an annulus from r to $r + dr$, therefore, would be proportional to $r^{-1}2\pi r\,dr$, and would be finite.)

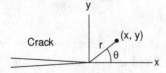

Figure 3.2. Crack-tip coordinate system.

The solution of the crack problem is assumed to be of the form

$$Z(z) = \frac{g(z)}{\sqrt{z}}$$

where $g(z)$ does not contain negative-power terms in z (that is, no z^{-n} terms). This solution satisfies the first of the two boundary conditions by virtue of its singular term; *i.e.*, $\sigma_{yy} \rightarrow \infty$ for $z \rightarrow 0$. To satisfy the second of the two conditions (*i.e.*, $\sigma_{yy} = 0$ for $x < 0, y = 0$), then

$$\sigma_{yy} = \Re e \left(\frac{g(z)}{\sqrt{x}} \right) = \Re e \left(\frac{g(x)}{\sqrt{x}} \right) = 0$$

This condition is satisfied only if $g(x)$ is real along $y = 0$ (because $\sqrt{x}$ is imaginary for $x < 0$). Taking a Taylor series expansion around the origin, the function $g(z)$ close to the origin then becomes

$$g(z) = g(0) + z \frac{dg(z)}{dz} + \frac{1}{2!} z^2 \frac{d^2 g(z)}{dz^2} + \cdots$$

Very close to the crack tip, therefore, $g(z) \approx g(0) = $ *real constant*. The constant is identified with the stress intensity factor K_I, and for convenience

$$g(0) \equiv \frac{K_I}{\sqrt{2\pi}}$$

Then,

$$Z(z) = \frac{K_I}{\sqrt{2\pi z}} \tag{3.28}$$

Expressing z in polar coordinates, $z = re^{i\theta}$, the function $Z(z)$ and its derivative $Z'(z)$ that are needed for the near-tip stresses may be written as follows:

$$Z(z) = \frac{K_I}{\sqrt{2\pi z}} = \frac{K_I}{\sqrt{2\pi r}} e^{-i\frac{\theta}{2}} = \frac{K_I}{\sqrt{2\pi r}} \left(\cos \frac{\theta}{2} - i \sin \frac{\theta}{2} \right)$$

$$Z'(z) = \frac{d}{dz} \frac{K_I}{\sqrt{2\pi z}} = -\frac{1}{2} \frac{K_I}{\sqrt{2\pi}} z^{-\frac{3}{2}} = -\frac{K_I}{\sqrt{2\pi r}} \frac{1}{2r} e^{-i\frac{3\theta}{2}}$$

$$= -\frac{K_I}{\sqrt{2\pi r}} \frac{1}{2r} \left(\cos \frac{3\theta}{2} - i \sin \frac{3\theta}{2} \right)$$

The real and imaginary parts are then:

$$\Re e\, Z(z) = \frac{K_I}{\sqrt{2\pi r}} \cos \frac{\theta}{2}$$

$$\Im m\, Z'(z) = \frac{K_I}{\sqrt{2\pi r}} \frac{1}{2r} \sin \frac{3\theta}{2}$$

$$\Re e\, Z'(z) = -\frac{K_I}{\sqrt{2\pi r}} \frac{1}{2r} \cos \frac{3\theta}{2}$$

By neglecting the higher-order terms, the stresses at the crack tip for tensile opening mode (mode I) loading are now given from Eqn. (3.24), by:

$$\sigma_{xx} = \Re e\, Z(z) - y\Im m\, Z'(z) = \frac{K_I}{\sqrt{2\pi r}}\cos\frac{\theta}{2}\left(1 - \sin\frac{\theta}{2}\sin\frac{3\theta}{2}\right)$$

$$\sigma_{yy} = \Re e\, Z(z) + y\Im m\, Z'(z) = \frac{K_I}{\sqrt{2\pi r}}\cos\frac{\theta}{2}\left(1 + \sin\frac{\theta}{2}\sin\frac{3\theta}{2}\right) \qquad (3.29)$$

$$\tau_{xy} = -y\Re e\, Z' = \frac{K_I}{\sqrt{2\pi r}}\cos\frac{\theta}{2}\sin\frac{\theta}{2}\cos\frac{3\theta}{2}$$

where

$$y = r\sin\theta = 2r\sin\frac{\theta}{2}\cos\frac{\theta}{2}$$

The stresses at the crack tip for forward shearing (mode II) and longitudinal shearing (mode III) modes of loading may be similarly obtained [1]. The stresses for mode II are given by Eqn. (3.30) as follows:

$$\sigma_{xx} = -\frac{K_{II}}{\sqrt{2\pi r}}\sin\frac{\theta}{2}\left(2 + \cos\frac{\theta}{2}\cos\frac{3\theta}{2}\right)$$

$$\sigma_{yy} = \frac{K_{II}}{\sqrt{2\pi r}}\sin\frac{\theta}{2}\cos\frac{\theta}{2}\cos\frac{3\theta}{2} \qquad (3.30)$$

$$\tau_{xy} = \frac{K_{II}}{\sqrt{2\pi r}}\cos\frac{\theta}{2}\left(1 - \sin\frac{\theta}{2}\sin\frac{3\theta}{2}\right)$$

Those for mode III are given in Eqn. (3.31) as follows:

$$\tau_{xz} = -\frac{K_{III}}{\sqrt{2\pi r}}\sin\frac{\theta}{2}$$

$$\tau_{yz} = \frac{K_{III}}{\sqrt{2\pi r}}\cos\frac{\theta}{2} \qquad (3.31)$$

The remaining task is to develop stress intensity solutions for specific crack and component geometries and loading conditions. For simple cases, closed-form solutions can be obtained. In many cases, stress intensity factors can be obtained through the use of the "method of superposition." A few simple cases are considered in the next section to illustrate the process for obtaining stress intensity factor solutions analytically. For more complex cases, the stress intensity factors may be obtained experimentally or numerically as described in Chapter 2 and references [2–4]. Solutions for many cases are available through handbooks [2–4] and are incorporated into a number of fracture analysis software programs. The stress intensity factors for some useful cases are given herein.

3.4 Stress Intensity Factors – Illustrative Examples

To illustrate the use of the Westergaard stress function approach, the case of a central crack of length 2a in an infinitely large thin plate (i.e., for generalized plane stress) subjected to two different loading conditions is considered. The solutions are then used to illustrate the method of superposition.

3.4.1 Central Crack in an Infinite Plate under Biaxial Tension (Griffith Problem)

The case of an infinitely large thin plate, containing a central through-thickness crack of length $2a$, subjected to remote, uniform biaxial tension is considered (Fig. 3.3). The boundary conditions are as follows:

$$\sigma_{yy} = \sigma \quad \text{and} \quad \tau_{xy} = 0 \quad \text{at} \quad y = \pm\infty$$
$$\sigma_{xx} = \sigma \quad \text{and} \quad \tau_{xy} = 0 \quad \text{at} \quad x = \pm\infty$$
$$\sigma_{yy} = 0 \quad \text{along} \quad y = 0; \quad -a \leq x \leq a \text{ (traction free along the crack surfaces)}$$

From symmetry, $\tau_{xy} = 0$ along the $y = 0$ plane.

Stress Intensity Factor

For this simple case, a solution may be obtained through examination of the boundary conditions. To satisfy the traction-free boundary condition along the crack surfaces and to account for stress intensification at the crack tip (*i.e.*, at $x = \pm a$), the stress function $Z(z)$ would need to be of the following form:

$$Z(z) \propto \frac{1}{\sqrt{z^2 - a^2}}$$

Specifically, because $Z(z)$ for $-a < x < a$ would be imaginary and $\sigma_{yy} = \Re e\, Z(z)$ along $y = 0$, Eqn. (3.24), σ_{yy} would be zero and satisfy the traction-free condition along the crack. At $x = \pm a$, the function would tend to infinity and, thereby, satisfy the required stress intensification. To additionally satisfy the remote traction boundary conditions, the following form of $Z(z)$ is chosen to be a possible solution:

$$Z(z) = \frac{\sigma z}{\sqrt{z^2 - a^2}} \tag{3.32}$$

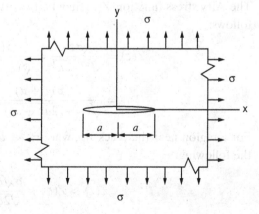

Figure 3.3. A central through-thickness crack in an infinitely large plate subjected to remote, uniform biaxial tension.

Equation (3.32) is differentiated with respect to z to obtain $Z'(z)$; *i.e.*,

$$Z'(z) = \frac{d}{dz}\left(\frac{\sigma z}{\sqrt{z^2 - a^2}}\right) = \frac{\sigma}{\sqrt{z^2 - a^2}} - \frac{\sigma z^2}{\left(\sqrt{z^2 - a^2}\right)^3} \tag{3.33}$$

Examination of Eqns. (3.32) and (3.33) shows that as $z \to \infty$:

$$Z(z) \to \sigma; \quad \text{and} \quad Z'(z) \to \frac{\sigma}{z} - \frac{\sigma}{z} = 0$$

Substitution into Eqn. (3.24) shows that:

As $z \to \infty$:
$$\sigma_{xx} = \Re e\, Z(z) - y\Im m\, Z'(z) \to \sigma$$
$$\sigma_{yy} = \Re e\, Z(z) + y\Im m\, Z'(z) \to \sigma$$
$$\tau_{xy} = -y\Re e\, Z'(z) \to 0$$
For $y = 0$:
$$\tau_{xy} = -y\Re e\, Z'(z) = 0$$

It is seen that, under this assumption, the stress function $Z(z)$ satisfies the boundary conditions and is a solution to the problem. The assumption of $A = 0$, however, needs to be verified further through a consideration of the displacements (see Eqn. (3.27)). For the moment, the assumption is deemed to be correct, and the process for obtaining the stress intensity factor is considered.

The stress intensity K_I is obtained by defining a new set of coordinates $\zeta = \xi + i\eta$ at the crack tip (see Fig. 3.3).

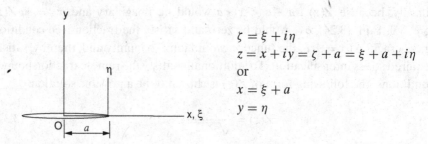

$$\zeta = \xi + i\eta$$
$$z = x + iy = \zeta + a = \xi + a + i\eta$$
or
$$x = \xi + a$$
$$y = \eta$$

The Airy stress function $Z(z)$ may be rewritten in terms of the new coordinates as follows:

$$Z(z) = \frac{\sigma z}{\sqrt{z^2 - a^2}} = \frac{\sigma(\zeta + a)}{\sqrt{(\zeta + a)^2 - a^2}}$$

$$= \frac{\sigma(\zeta + a)}{\sqrt{(\zeta^2 + 2a\zeta + a^2) - a^2}} = \frac{\sigma(\zeta + a)}{\sqrt{\zeta^2 + 2a\zeta}}$$

For a region near the crack tip, where $\zeta \ll a$, then $Z(z)$ is given approximately by the following:

$$Z(z) \approx Z(\zeta) \approx \frac{\sigma\sqrt{a}}{\sqrt{2\zeta}} = \frac{\sigma\sqrt{\pi a}}{\sqrt{2\pi\zeta}} \tag{3.34}$$

By comparing Eqns. (3.34) and (3.28), it is readily seen that the mode I stress intensity factor may be defined by:

$$K_I \equiv \lim_{\zeta \to 0} \sqrt{2\pi\zeta}\, Z(\zeta) = \sigma\sqrt{\pi a} \tag{3.35}$$

This is in agreement with that was obtained through Griffith's energy consideration in Chapter 2.

Displacements

To check on the assumption of $A = 0$, the displacements for generalized plane stress are considered. Using Hooke's law and the definition of strains in terms of displacements, the displacements $u(x, y)$ and $v(x, y)$ may be obtained by integration as follows:

$$Eu = (1 - v)\Re e\,\overline{Z}(z) - (1 + v)y\Im m\,Z(z) + (constant)_1 + f_1(y) \tag{3.36}$$

$$Ev = 2\Im m\,\overline{Z}(z) - (1 + v)y\Re e\,Z(z) + (constant)_2 + f_2(x) \tag{3.37}$$

The constants in Eqns. (3.36) and (3.37) represent rigid-body translation and may be disregarded in considering deformation within the body. The functions $f_1(y)$ and $f_2(x)$ may be examined through a consideration of the shearing strain γ_{xy}. From Eqns. (3.6), (3.7), and (3.24),

$$E\gamma_{xy} = E\left(\frac{\partial u}{\partial y} + \frac{\partial v}{\partial x}\right) = -2(1 + v)y\Re e\,Z'(z) + \frac{\partial f_1(y)}{\partial y} + \frac{\partial f_2(x)}{\partial x} \equiv 2(1 + v)\tau_{xy} \tag{3.38}$$

It is clear then that the sum of the two derivatives in Eqn. (3.26) or (3.38) must be zero; *i.e.*,

$$\frac{\partial f_1(y)}{\partial y} + \frac{\partial f_2(x)}{\partial x} = -A + A = 0$$

In other words, $f_1(y) = -Ay$ and $f_1(x) = Ax$, which correspond to rigid body rotation about the z-axis. The sign is chosen to be consistent with a counter-clockwise rotation. It is clear that the constant A could not be arbitrarily neglected; it is zero only for the case of equal biaxial tension.

3.4.2 Central Crack in an Infinite Plate under a Pair of Concentrated Forces [2–4]

Wedge force loading applied normally to the crack plane often occurs in many practical applications. The loading is illustrated in Fig. 3.4 for a pair of concentrated forces P (force per unit thickness).

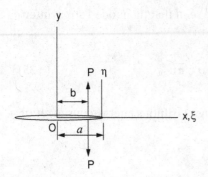

Figure 3.4. A central through-thickness crack in an infinitely large plate subjected to a pair of concentrated forces applied normal to the crack surface.

Based on the solution of concentrated forces in mode III loading [2], the following Airy's stress function is assumed:

$$Z(z) = \frac{P\sqrt{b^2 - a^2}}{i\pi(z-b)\sqrt{z^2 - a^2}} = \frac{P\sqrt{a^2 - b^2}}{\pi(z-b)\sqrt{z^2 - a^2}} \qquad (3.39)$$

The function satisfies the boundary conditions and accounts for the impact of off-center loading on the two crack tips; *i.e.*:

$$\frac{1}{\sqrt{z^2 - a^2}} \text{ guarantees that } \begin{cases} \sigma_{yy} = 0 \quad \text{for } -a < x < +a, \, y = 0 \\ \sigma_{yy} = \sigma_{xx} = \tau_{xy} = 0 \quad \text{as } z \to \infty \end{cases}$$

$\frac{\sqrt{a^2-b^2}}{z-b}$ as a whole accounts for the effect of off-center loading at each crack tip

$$\text{For } z = +a \frac{\sqrt{a^2 - b^2}}{z - b} \to \frac{\sqrt{(a+b)(a-b)}}{a - b} = \sqrt{\frac{a+b}{a-b}}$$

$$\text{For } z = -a \frac{\sqrt{a^2 - b^2}}{z - b} \to \frac{\sqrt{(a+b)(a-b)}}{-(a+b)} = -\sqrt{\frac{a-b}{a+b}}$$

In other words, the stresses would be higher at the right end ($x = +a$) relative to the left end ($x = -a$).

Again, by taking $z = \zeta + a$ for the right end, the Airy stress function in terms of the crack-tip coordinates is given by:

$$Z_R(z) = \frac{P\sqrt{a^2 - b^2}}{\pi(z-b)\sqrt{z^2 - a^2}} = \frac{P\sqrt{a^2 - b^2}}{\pi(\zeta + a - b)\sqrt{(\zeta + a)^2 - a^2}} \approx \frac{P}{\pi\sqrt{2\zeta a}}\sqrt{\frac{a+b}{a-b}}$$

The crack-tip stress intensity factor is given as follows:

$$K_R(z) = \lim_{\zeta \to 0} \sqrt{2\pi\zeta} \, Z_R(z) = \frac{P}{\sqrt{\pi a}}\sqrt{\frac{a+b}{a-b}} \qquad (3.40)$$

By moving the crack-tip coordinates to the left end, then, $z = -\zeta - a$. The same procedure leads to:

$$Z_L(z) = \frac{P\sqrt{a^2 - b^2}}{\pi(z-b)\sqrt{z^2 - a^2}} = \frac{P\sqrt{a^2 - b^2}}{\pi(-\zeta - a - b)\sqrt{(-\zeta - a)^2 - a^2}} \approx -\frac{P}{\pi\sqrt{2\zeta a}}\sqrt{\frac{a-b}{a+b}}$$

The stress intensity factor at the left tip is then:

$$K_L(z) = \lim_{\zeta \to 0} \left(-\sqrt{2\pi\zeta}\, Z_L(z) \right) = \frac{P}{\sqrt{\pi a}} \sqrt{\frac{a-b}{a+b}} \tag{3.41}$$

3.4.3 Central Crack in an Infinite Plate under Two Pairs of Concentrated Forces

Because the solutions are based on linear elasticity, the solution for a central crack in an infinite plate under two pairs of concentrated forces P (force per unit thickness), placed symmetrically about the center line (see Fig. 3.5), may be obtained by simple superposition of Eqns. (3.40) and (3.41).

Namely,

$$K_I = K_L + K_R = \frac{P}{\sqrt{\pi a}} \sqrt{\frac{a-b}{a+b}} + \frac{P}{\sqrt{\pi a}} \sqrt{\frac{a+b}{a-b}} = \frac{2Pa}{\sqrt{\pi a(a^2 - b^2)}} \tag{3.42}$$

Equation (3.42) may be used as a Green's function to develop stress intensity factor solutions for other loading conditions.

3.4.4 Central Crack in an Infinite Plate Subjected to Uniformly Distributed Pressure on Crack Surfaces

This example is used to illustrate the direct application of the superposition method and the use of the Green's function approach. For the superposition method, the loading is viewed as being the difference between a cracked body under uniform remote traction σ and one in which the crack is held shut by uniform traction σ along the crack surfaces, with the magnitude of the traction equal to the pressure p along the crack (see Fig. 3.6).

The solution for the first case is known. The second case corresponds to that of an uncracked plate for which the stress intensification would be zero. The stress intensity factor for the pressure-loaded crack, therefore, would be equal to that of the remotely loaded crack; namely,

$$K_I = K_a - K_b = \sigma\sqrt{\pi a} - 0 = p\sqrt{\pi a} \tag{3.43}$$

Figure 3.5. A central through-thickness crack in an infinitely large plate subjected to two pairs of concentrated forces applied normal to the crack surface.

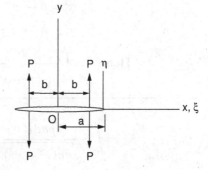

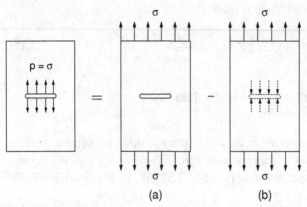

(a)   (b)

Figure 3.6. A central through-thickness crack in an infinite plate subjected to uniformly distributed pressure on the crack surfaces.

The solution may be obtained also by using the results for two pairs of concentrated forces (Section 3.4.3) as a Green's function. The distance of the force from the center b is taken to be a variable x' and the concentrated force P is replaced by the incremental contribution pdx'. The contribution to the stress intensity factor is given from Eqn. (3.42) as:

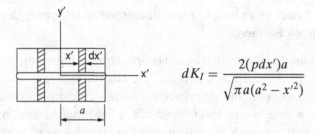

$$dK_I = \frac{2(pdx')a}{\sqrt{\pi a(a^2 - x'^2)}}$$

The stress intensity factor is then obtained by integrating over the interval $x' = 0$ to $x' = a$, because the Green's function is based on two pairs of forces.

Integration is made by changing the variable from x' to u.

Let $\qquad x' = a \sin u$

Then $\qquad dx' = a \cos u du$

$\qquad\qquad x' = 0 \rightarrow u = 0$

$\qquad\qquad x' = a \rightarrow u = \dfrac{\pi}{2}$

Hence, $\qquad \displaystyle\int_0^a \frac{dx'}{\sqrt{(a^2 - x'^2)}} = \int_0^{\frac{\pi}{2}} \frac{a \cos u du}{\sqrt{(a^2 - a^2 \sin^2 u)}} = \frac{\pi}{2}$

$$K_I = \int_0^a \frac{2padx'}{\sqrt{\pi a(a^2 - x'^2)}} = \frac{2pa}{\sqrt{\pi a}} \int_0^a \frac{dx'}{\sqrt{(a^2 - x'^2)}} = p\sqrt{\pi a} \qquad (3.44)$$

This example illustrates the validity and utility of the different approaches for using known solutions.

3.5 Relationship between *G* and *K*

Having derived expressions for the stresses and displacements near the crack tip (see Sections 3.3 and 3.4), it is now possible to formally consider the relationship between the strain energy release rate *G* and the stress intensity factor *K* based on the first law of thermodynamics. The first law states that the work done on a system is equal to the increase in internal energy of the system, *i.e.*,

$$Work\ done = \Delta(internal\ energy)$$

Note that the work done is associated only with the change in stored elastic energy in the system and, as such, implicitly assumes *isothermal conditions*.

The case of a cracked body under fixed displacement loading is considered, and attention is focused on the crack-tip region. Referring to Fig. 3.7, the material is first slit open along the crack plane (or the *x* direction) by an amount α, which is then maintained shut by the imposition of traction that is equal to the crack-tip stresses σ_{yy}. The traction (viewed as external forces) is then allowed to relax to zero so that the crack now opens to the stress-free opening *v*. According to the first law, the work done in this stress relaxation is equal to the change in strain energy in the cracked body.

The work done on the system is given by the stress σ_{yy} and displacement *v* along $y = 0$ and the change in strain energy is given as $(dU/dA)_\Delta \Delta A$ over the increment of crack extension of area ΔA, where dA represents an elemental area over which the stress acts. According to the first law, then:

$$2 \int_{\Delta A} \frac{1}{2}(-\sigma_{yy}dA) \times v = \left(\frac{dU}{dA}\right)_\Delta \Delta A \tag{3.45}$$

The minus sign in the work term reflects the fact that the stress (σ_{yy}) and displacement (*v*) act in opposite directions during the relaxation, and the factor 2 accounts for action on both crack surfaces. Because the strain energy release rate *G* is defined

Figure 3.7. Stresses and displacements near the crack tip.

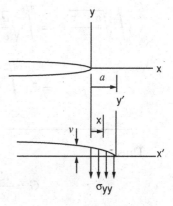

as $-(dU/dA)_\Delta$ for fixed displacement loading (see Eqn. (2.25)), then:

$$G = \lim_{\Delta A \to 0} \frac{2}{\Delta A} \int_{\Delta A} \frac{1}{2} \sigma_{yy} v \, dA$$

or,

$$G = \lim_{\alpha \to 0} \frac{2}{\alpha} \int_0^\alpha \frac{1}{2} \sigma_{yy} v \, dx \quad \text{for a plate of unit thickness} \tag{3.46}$$

From Eqn. (3.28), the Airy stress function $Z(z)$ is given by:

$$Z(z) = \frac{K_I}{\sqrt{2\pi z}}$$

The relevant stress and displacement at the crack tip are then as follows:

$$\sigma_{yy} = \Re e\, Z + y \Im m\, Z'(z) = \Re e\, Z(z) = \frac{K_I}{\sqrt{2\pi x}} \quad \text{for } x > 0,\, y = 0$$

$$v = \frac{1}{E}\left(2\Im m\, \overline{Z}(z) - (1+v)y\Re e\, Z(z)\right) = \frac{2}{E}\Im m\, \overline{Z}(z) = \frac{2}{E}\Im m\left(\frac{2K_I\sqrt{z}}{\sqrt{2\pi}}\right) \quad \text{along } y = 0$$

The stress of interest is referenced to the current crack tip. The displacement of interest, on the other hand, is behind the crack tip, over the cut length α that is to be released, and is to be referenced to the new tip at $x = \alpha$, or at $x' = -(x - \alpha)$; or $z = -(x - \alpha) + iy$, $y = 0$ (see Fig. 3.7). Thus,

$$v = \frac{2}{E}\Im m\left(\frac{2K_I\sqrt{-(\alpha - x)}}{\sqrt{2\pi}}\right) = \frac{4K_I\sqrt{\alpha - x}}{E\sqrt{2\pi}}$$

Substitution into Eqn. (3.46) then gives:

$$G = \lim_{\alpha \to 0} \frac{2}{\alpha} \int_0^\alpha \frac{1}{2}\left(\frac{K_I}{\sqrt{2\pi x}}\right)\left(\frac{4K_I\sqrt{\alpha - x}}{E\sqrt{2\pi}}\right) dx = \lim_{\alpha \to 0} \frac{2K_I^2}{\alpha \pi E} \int_0^\alpha \sqrt{\frac{\alpha - x}{x}} \, dx$$

Integration is made again by substitution of variables.

Let $x = \alpha \sin^2 u$　　　then　$dx = 2\alpha \sin u \cos u\, du$

$$\int_0^\alpha \sqrt{\frac{\alpha - x}{x}} \, dx = \int_0^{\pi/2} \sqrt{\frac{\alpha}{\alpha}}\sqrt{\frac{1 - \sin^2 u}{\sin^2 u}} 2\alpha \sin u \cos u\, du$$

$$= 2\alpha \int_0^{\pi/2} \frac{\cos u}{\sin u} \sin u \cos u\, du = \alpha \int_0^{\pi/2} (1 + \cos 2u)\, du = \frac{\pi}{2}\alpha$$

Thus,

$$G = \frac{K_I^2}{E} \quad \text{for generalized plane stress} \tag{3.47a}$$

Similarly, one can show that

$$G = \frac{(1 - v^2)K_I^2}{E} \text{ for plane strain} \tag{3.47b}$$

The engineering community prefers to work with stresses rather than energy. As such, stress intensity factors K (in units of $(F/L^2)(L^{1/2})$) is now widely used in engineering, whereas the use of strain energy release rate G (in units of (FL/L^2)) is limited to some scientific areas.

3.6 Plastic Zone Correction Factor and Crack-Opening Displacement

Before closing this chapter, two plasticity-related parameters need to be introduced. The first parameter relates to the presence of plastic deformation at the crack tip in technologically important material (*i.e.*, the plastic zone correction factor), and an estimation of its size. The second one relates to the extent of opening of the crack at its tip in the presence of plastic deformation, which is then used as an alternate parameter for characterizing the crack-driving force.

Plastic Zone Correction Factor

An estimate of the plastic zone correction factor was made by Irwin (see [9]). He postulated that stresses ahead of the crack tip, away from a "small" plastic zone, can be approximated by those given by the solutions of linear elasticity, provided that an effective crack length a_e is used, where,

$$a_e = a + r_y \tag{3.48}$$

In Eqn. (3.48), a is the physical (actual) crack length, and r_y is the plastic zone correction factor. The parameter r_y, for generalized plane stress, is estimated by setting the normal stress $\sigma_{yy}(r, 0)$ directly in front of the crack equal to the uniaxial tensile yield strength σ_{ys} of the material. From Eqn. (3.29), one obtains:

$$\sigma_{yy}(r_y, 0) = \frac{K}{\sqrt{2\pi r_y}} = \sigma_{ys}$$

The plastic zone correction factor, for generalized plane stress, is, therefore, given by Eqn. (3.49) below.

$$r_y = \frac{1}{2\pi} \left(\frac{K}{\sigma_{ys}}\right)^2 \tag{3.49}$$

Because of the constraint imposed under plane strain conditions, yielding (onset of plastic deformation) would occur at a higher stress level. A number of estimates were made with different assumed constraint and yielding criteria [9]. But, because of the approximate nature of these estimates, the plastic zone correction factor for plane strain is taken to be that given by Eqn. (3.50).

$$r_{Iy} = \frac{1}{6\pi} \left(\frac{K_I}{\sigma_{ys}}\right)^2 \tag{3.50}$$

The convention of retaining and omitting the subscript I to differentiate between plane strain and generalized plane stress (or nonplane strain) loading, respectively, for mode I loading is adopted for Eqns. (3.49) and (3.50).

As a result of stress redistribution due to yielding, plastic deformation is expected to extend further ahead of the crack tip than that indicated by the plastic zone correction factors. For simplicity, and to an acceptable degree of accuracy for engineering analyses, the plastic zone size is taken to be equal to twice the plastic zone correction factor, i.e.,

$$r_p \approx 2r_y = \frac{1}{\pi}\left(\frac{K}{\sigma_{ys}}\right)^2 \quad \text{for plane stress}$$

$$r_{Ip} \approx 2r_{Iy} = \frac{1}{3\pi}\left(\frac{K_I}{\sigma_{ys}}\right)^2 \quad \text{for plane strain} \tag{3.51}$$

Crack-Tip-Opening Displacement (CTOD)

The crack-opening displacement (COD) at any point along the crack is defined as twice the displacement of the crack surface at that location, i.e., $COD \equiv 2v(-r,0)$. The quantity crack-tip-opening displacement (CTOD), however, is given a special designation as the opening displacement at the actual crack tip, which is assumed to be correctly located at a distance $x = -r_y$ from the effective crack tip based on Irwin's approximation. As such, from Eqns. (3.28), (3.37), and (3.49), one obtains:

$$v(-r_y,0) = \frac{2}{E}\Im m\,\overline{Z}(z) = \frac{4K\sqrt{r_y}}{\sqrt{2\pi}\,E} = \frac{4K}{\sqrt{2\pi}\,E}\sqrt{\frac{1}{2\pi}\left(\frac{K}{\sigma_{ys}}\right)^2} = \frac{2K^2}{\pi E\sigma_{ys}}$$

The CTOD is give, therefore, by Eqn. (3.52) below.

$$CTOD = 2v(-r_y,0) = \frac{4K^2}{\pi E\sigma_{ys}} = \frac{4G}{\pi\sigma_{ys}} \tag{3.52}$$

The CTOD at fracture is used sometimes, particularly in England, as a fracture criterion for low-strength alloys [10, 11].

3.7 Closing Comments

A brief summary of the linear elasticity framework for fracture mechanics (i.e., the analysis of cracked bodies) is presented, and the use of the principle of superposition to obtain solutions for more complex loading configurations is introduced. A simple (consensus) "correction factor" to account for plastic deformation at the crack tip is identified. The readers are encouraged to consult published literature to gain a broader overview of this area, and to access solutions for other crack and loading configurations. The remaining chapters will be devoted to the application of linear fracture mechanics to the study of fracture and crack growth and their application in relation to structural integrity and durability.

REFERENCES

[1] Irwin, G. R., "Analysis of Stresses and Strains Near the End of a Crack Traversing a Plate," J. Applied Mechanics, ASME, 24 (1957), 361.

[2] Sih, G. C., ed., "Methods of Analysis and Solutions of Crack Problems," Mechanics of Fracture 1, Noordhoff Int'l. Publ., Leyden, The Netherlands (1973).

[3] Tada, H., Paris, P. C., and Irwin, G. R., "The Stress Analysis of Cracks Handbook," 3rd ed., ASME Press, New York (2000).

[4] Broek, D., in "Elementary Engineering Fracture Mechanics," 4th ed., Martinus Nijhoff Publishers, Leiden, The Netherlands (1986).

[5] Mushkilishevili, N., "Some Basic Problems of the Mathematical Theory of Elasticity," 4th corrected and augmented edition, Moscow 1954, Translated by J. R. M. Radok, P. Noordhoff, Groningen, The Netherlands (1963).

[6] Sokolnikoff, I. S., "Mathematical Theory of Elasticity," 2nd ed., McGraw-Hill Book Co., Inc., New York (1956).

[7] Timoshenko, S. "Theory of Elasticity," 2nd ed., McGraw-Hill Book Co., Inc., New York (1951).

[8] Westergaard, H. M., "Bearing Pressures and Cracks," J. Appl. Mech., 61 (1939), A49–A53.

[9] Brown, W. F. Jr., and Srawley, J. E., "Plane Strain Crack Toughness Testing of High Strength Metallic Materials," ASTM Special Technical Publication 410, American Society for Testing and Materials and National Aeronautics and Space Administration (1965).

[10] Wells, A. A., "Unstable Crack Propagation in Metals-Cleavage and Fast Fracture," Proc. Crack Propagation Symposium, Cranfield (1961), 210–230.

[11] Wells, A. A., "Application of Fracture Mechanics at and Beyond General Yielding," British Welding Research Assoc. Report M13 (1963).

4 Experimental Determination of Fracture Toughness

In the preceeding chapters, the physical basis and analytical framework, based on linear elasticity, for addressing the issue of unstable or sudden fracture of engineering materials were presented. The driving force for fracture, or crack growth, is characterized in terms of the strain energy release rate G, or the crack-tip stress intensity factor K defined through the linear elasticity analysis. Crack growth instability, or sudden fracture, would occur when these parameters reached their "critical" values. These values represent the material property conjugate to the crack-driving forces (G or K), *i.e.*, the fracture toughness. With the present state of understanding, fracture toughness cannot be calculated based on other mechanical properties and must be measured experimentally. Because the underlying analytical framework is that of linear elasticity, and the materials of engineering interest are expected to undergo nonelastic deformations at the crack tip, measurements of fracture toughness and the utilization of this information in design must conform to conditions under which linear elastic analysis can serve as a "good" approximation. In this chapter, the experimental procedures for determining fracture toughness are described. The analytical and empirical bases for the design of specimens and the interpretation of test records are summarized. Before discussing the methods for measuring fracture toughness, it is important to first examine the consequences of plastic deformation at the crack tip in relation to fracture.

4.1 Plastic Zone and Effect of Constraint

In Chapters 2 and 3, the restrictions in the use of linear elastic fracture mechanics (LEFM) were discussed in terms of the dimensions of the crack and the body (specimen, component, or structure) relative to the size of the crack-tip plastic zone. Simple estimates of the plastic zone sizes were given in Section 3.6. A more detailed examination of the role of constraint (plane strain versus plane stress) and the variations in plastic zone size from the surface to the interior of a body would help in understanding fracture behavior and the design of practical specimens for measurements of fracture toughness. Note that the plastic zone size in actual materials is a function of its deformation characteristics (*i.e.*, its stress-strain or constitutive behavior) and cannot be readily calculated. For the purposes of this chapter, it is sufficient to develop a semiquantitative appreciation of its implications in terms of an elastic-perfectly plastic material through the use of one of the classical criteria for yielding (or the onset of plastic deformation). Focus will be placed on mode I loading.

For mode I loading, the stresses near the tip of a crack in a plate are given from Eqn. (3.29) in polar-cylindrical coordinates, with the z-axis along the crack front, the x-axis in the direction of crack prolongation, and the y-axis perpendicular to the crack plane.

$$\sigma_{xx} = \frac{K_I}{\sqrt{2\pi r}} \cos\frac{\theta}{2} \left(1 - \sin\frac{\theta}{2} \sin\frac{3\theta}{2}\right)$$

$$\sigma_{yy} = \frac{K_I}{\sqrt{2\pi r}} \cos\frac{\theta}{2} \left(1 + \sin\frac{\theta}{2} \sin\frac{3\theta}{2}\right)$$

$$\sigma_{zz} = \begin{cases} 0 & \text{for plane stress} \\ v(\sigma_{xx} + \sigma_{yy}) & \text{for plane strain} \end{cases} \tag{4.1}$$

$$\tau_{xy} = \frac{K_I}{\sqrt{2\pi r}} \cos\frac{\theta}{2} \sin\frac{\theta}{2} \cos\frac{3\theta}{2}$$

$$\tau_{yz} = \tau_{zx} = 0$$

For convenience, the von Mises criterion for yielding is selected and is given by Eqn. (4.2) (see Eqn. (2.8)):

$$\left[(\sigma_{xx} - \sigma_{yy})^2 + (\sigma_{yy} - \sigma_{zz})^2 + (\sigma_{zz} - \sigma_{xx})^2\right] + 6\left[\tau_{xy}^2 + \tau_{yz}^2 + \tau_{zx}^2\right]$$
$$= \left[(\sigma_1 - \sigma_2)^2 + (\sigma_2 - \sigma_3)^2 + (\sigma_3 - \sigma_1)^2\right] = 2\sigma_{YS}^2 \tag{4.2}$$

The locus of yielding r_{ep} (*i.e.*, the boundary between yielded and elastic region) as a function of θ is obtained by substituting Eqn. (4.1) directly into Eqn. (4.2), or by transforming Eqn. (4.1) into principal stresses first, collecting terms, and simplifying through trigonometric identities. For this estimate, redistribution of stresses that results from yielding (or plastic deformation) near the crack tip is not considered. The yield loci, for plane stress and plane strain conditions, are given by Eqns. (4.3) and (4.4), respectively.

$$r_{ep} = \left(\frac{K_I^2}{2\pi\sigma_{YS}^2}\right) \left\{\frac{1}{2}\left[1 + \cos\theta + \frac{3}{2}\sin^2\theta\right]\right\} \text{ for plane stress} \tag{4.3}$$

$$r_{ep} = \left(\frac{K_I^2}{2\pi\sigma_{YS}^2}\right) \left\{\frac{1}{2}\left[(1 - 2v)^2(1 + \cos\theta) + \frac{3}{2}\sin^2\theta\right]\right\} \text{ for plane strain} \tag{4.4}$$

Equations (4.3) and (4.4) are written to expressly reflect Irwin's plastic zone correction factor r_y (see Eqn. (3.49)). By normalizing with respect to r_y, the yield loci, or

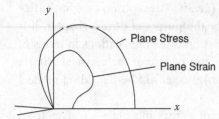

Figure 4.1. Estimated plastic zone sizes based on von Mises criterion for yielding ($v = 0.3$ for the plane strain case).

plastic zone sizes, are given in nondimensional form by Eqns. (4.5) and (4.6).

$$\frac{r_{ep}}{r_y} = \frac{r_{ep}}{\left(\dfrac{K_I^2}{2\pi\sigma_{YS}^2}\right)} = \frac{1}{2}\left[1 + \cos\theta + \frac{3}{2}\sin^2\theta\right] \text{ for plane stress} \qquad (4.5)$$

$$\frac{r_{ep}}{r_y} = \frac{r}{\left(\dfrac{K_I^2}{2\pi\sigma_{YS}^2}\right)} = \frac{1}{2}\left[(1 - 2v)^2(1 + \cos\theta) + \frac{3}{2}\sin^2\theta\right] \text{ for plane strain} \qquad (4.6)$$

The size and shape of the plastic zones are illustrated in Fig. 4.1; the plane strain zone is estimated for a Poisson ratio v of 0.3. Note that the actual zone sizes would be larger to reflect redistribution of stresses associated with plastic deformation near the crack tip. The shape of the plastic zone would also change to reflect the work-hardening behavior of the material. The actual sizes and shapes would need to be determined experimentally for each class of materials.

The key points to be gleaned from this exercise are that the plastic zone size depends on the state of stress (or constraint) and is proportional to $(K_I/\sigma_{ys})^2$. Its size is expected to vary through the thickness of a plate, and it would increase with increasing stress intensity factor K_I and decreasing yield strength σ_{ys}. The consequence of plastic deformation on fracture behavior and fracture toughness measurements is considered briefly in the next section.

4.2 Effect of Thickness; Plane Strain versus Plane Stress

It is now possible to take a more detailed look at the consequences of plastic deformation at the crack tip. Because of the high stress gradient at the crack tip (see Eqn. (4.1)), the material nearer to the crack tip would undergo greater lateral (Poisson) contraction than materials that are further away. The elastic material away from the crack tip, therefore, would exert constraint (*i.e.*, restrict lateral contraction) on the near-tip material. The degree of constraint would vary through the thickness of the material: From plane stress at and near the surface and tending toward plane strain in the interior. The extent of near-tip plastic deformation would reflect this through-thickness variation in constraint and the constraint, in turn, would be reduced by this deformation. A schematic representation of the variation in plastic zone size, based on the von Mises criterion, is shown in Fig. 4.2. The effectiveness of the constraint is expected to depend on the ratio of the material

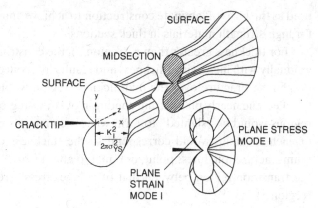

Figure 4.2. Schematic representation of the through-thickness variation in plastic zone size based on the von Mises criterion for yielding [1].

thickness (B) and some measure of the plastic zone size, for example, Irwin's plastic zone correction factor r_y. Namely, $B/r_y = B/(K_I/\sigma_{ys})^2$.

Because crack-tip plastic deformation would accompany crack growth, the work of deformation would contribute to the work of crack growth, or the fracture toughness (K_c or G_c) of the material. As such, K_c or G_c would vary as a function of thickness to reflect the changing constraint on crack-tip plastic deformation. Indeed, experimental observations show that the typical variation in fracture toughness with thickness (or B/r_y) for a single material would be shown by the schematic diagram in Fig. 4.3.

The diagram may be divided into three regions: (1) one where B is less than or equal to r_y, (2) one in which B is larger than, but is of the order of r_y; and (3) where B is much larger than r_y. In region 1, or the "plane stress" region, the relief of constraint is essentially complete and the stress state at the crack tip approximates that of plane stress. The plane stress plastic zones from each surface tend to merge and fracture tends to occur by macroscopic shearing along the "elastic-plastic" interface to produce a "slant" fracture (or combined mode I and mode III fracture). Because the extent of plastic deformation would be limited to the order of the material thickness, the measured fracture toughness (K_c or G_c) will decrease with thickness. The maximum fracture toughness for a material tends to occur at $B \approx r_y$. This behavior is

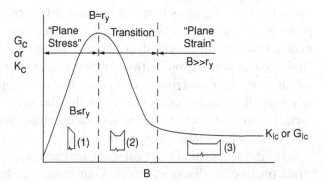

Figure 4.3. Schematic diagram showing the typical variation in fracture toughness with material thickness (B), or thickness relative to Irwin's plastic zone correction factor (B/r_y).

used as the basis of laminate construction to achieve "maximum" fracture toughness for high-strength materials in thick sections.

For thickness greater than r_y (region 2), the constraint at the crack tip increases gradually with increasing thickness and results in a concomitant decrease in fracture toughness. Fracture over the midthickness region would be macroscopically "flat" to reflect the nearly plane strain condition over this region, whereas the near-surface regions would be "slanted" to form what is commonly called "shear lips." The size of each shear lip would correspond to the thickness of the material at its maximum fracture toughness point, or equal to about $0.5r_y$. As such, region 2 represents the transition region between that of "plane stress" (region 1) and "plane strain" (region 3).

For thickness much greater than r_y, the crack-tip constraint is at its maximum and approximates that of plane strain. The approximate nature of plane strain arises from the fact that there is no lateral constraint at the surface. As such, the surface region would always be under the state of plane stress. When the thickness is large, however, this plane stress region would be small relative to the predominantly plane strain region along the crack front in the interior. The fracture toughness would be at its minimum. This so-called plane strain fracture toughness (K_{Ic} or G_{Ic}) is considered to be the intrinsic fracture toughness of the material, and is used as the basis for material development and structural integrity analyses.

From this brief discussion, it should be clear from the mechanics perspective that the fracture toughness of a material would reflect its yield strength and its thickness, in addition to the inherent toughness provided by its microstructure. Because of these influences, the design of specimens to properly measure fracture toughness (typically not known beforehand) is not straightforward. In the following sections, the methodologies for fracture toughness testing are discussed to provide an appreciation of the processes that are involved in arriving at standard methods and the associated testing and data analysis procedures.

4.3 Plane Strain Fracture Toughness Testing

Considerable focus is placed on the determination of plane strain fracture toughness, and well-defined international standard methods of test (*e.g.*, American Society of Testing and Materials (ASTM) Method E-399) are available. Interest in plane strain fracture toughness is based on several factors that were mentioned in the previous section. First, it is considered to be the intrinsic fracture toughness of a material. Because of the well-defined stress state in its determination (namely, plane strain), direct comparisons between different materials can be made. As such, it is suitable for use in material selection and alloy development. Because it is believed to control the onset of crack growth, in the absence of environmental effects and cyclic loading (fatigue), it is of interest for durability and structural integrity analyses, particularly for internal cracks that may be present or develop in large sections.

The development of practical specimens and procedures for determining plane strain fracture toughness was carried out during the late 1950s and 1960s, largely through the cooperative efforts of many researchers in the government, industry, and academe. The focal point of this activity was a special committee of the American Society of Testing and Materials (ASTM) which later became ASTM Committee on Fracture Testing of Metallic Materials, and now Committee E-08 on Fracture and Fatigue. The following discussion closely parallels the development of this group, which is reflected in ASTM Special Technical Publication (STP) 410 on "Plane Strain Fracture Toughness Testing of Metallic Materials" [1].

4.3.1 Fundamentals of Specimen Design and Testing

To understand the important factors in the design of practical specimens for plane strain fracture toughness (K_{Ic} or G_{Ic}) measurements, it is useful to begin by considering a configuration that is as simple as possible. The simplest configuration is that of an axially symmetric, circular (or penny-shaped) crack embedded inside a sufficiently large body so that the influences of its external boundary surface on the stress field of the crack are negligible (Fig. 4.4).

Initially (*i.e.*, before any load is applied to the body), the crack is regarded as being ideally sharp and is free from any self-equilibrating stresses (namely, residual stresses). The residual stresses might be those that result from the effects of generating the crack by fatigue loading in a practical test specimen, for example. This "specimen" is tested by steadily increasing the remotely applied (gross section) tensile stress, σ.

The mode I (tensile-opening mode) stress intensity factor at every point along the crack border is given by Eqn. (4.7).

$$K_I = 2\sigma \left(\frac{a}{\pi}\right)^{\frac{1}{2}} \tag{4.7}$$

In Eqn. (4.7), $2a$ is the effective crack diameter. Formally $2a$ represents the diameter of the physical crack and the associated plastic zone correction factor, namely,

$$2a = 2a_o + 2r_{Iy} = 2a_o + 2\left(\frac{1}{6\pi}\frac{K_I^2}{\sigma_{ys}^2}\right) = 2a_o + \frac{1}{3\pi}\frac{K_I^2}{\sigma_{ys}^2} \tag{4.8}$$
$$2a \approx 2a_o \quad \text{when } \sigma \ll \sigma_{ys}$$

To conduct a satisfactory K_{Ic} measurement, it is necessary to provide for autographic recording of the applied stress (or load) versus the output of a transducer

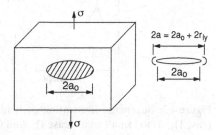

Figure 4.4. Schematic diagram of a circular (penny-shaped) crack inside a large body, subjected to uniformly applied, remote tensile stress perpendicular to the crack plane.

that accurately senses some quantity that can be related to the extension of the crack. The basic measurement, for this purpose, is the *displacement* of two points located symmetrically on opposite sides of the crack plane (see Fig. 4.4). For this hypothetical specimen, with an internal penny-shaped crack, this displacement can be measured only in principle.

If there is no crack growth during loading (*i.e.*, with the effective crack diameter, $2a$, remaining constant), the slope of the load-displacement trace will remain constant. If $2a$ increases, on the other hand, the slope will decrease. This decrease in slope would be associated with either actual crack extension or the development and growth of a plastically deformed zone at the crack tip (*i.e.*, apparent or effective crack extension) or both. The change in slope can be abrupt to reflect a sudden burst of crack extension. As such, the load-displacement record provides an effective means for assessing the specimen behavior and identifying the onset of crack growth.

For this penny-shaped crack model specimen, the crack diameter would be the only dimension of concern; the other dimensions would be taken to be very large. The crack size requirement may be considered by writing the effective crack diameter $2a$ in terms of the plane strain fracture toughness K_{Ic} and the yield strength σ_{ys} of the material by using Eqn. (4.7); i.e., for the conceptual case where yielding and fracture occur concurrently.

$$2a^* = \frac{\pi}{2} \left(\frac{K_{Ic}}{\sigma_{ys}} \right)^2 \approx 1.5 \left(\frac{K_{Ic}}{\sigma_{ys}} \right)^2 \tag{4.9}$$

In essence, the physical crack size $2a_o$ is being considered in relation to the crack-tip plastic zone size through the following three cases:

CASE I. $2a_o \gg (K_{Ic}/\sigma_{ys})^2$, where $(K_{Ic}/\sigma_{ys})^2 \propto r_{Iy}$ (overly large a_o).

In this case, the crack size is much larger than the plane strain crack-tip plastic zone size. As such the effective crack length $2a = 2a_o + 2r_{Iy}$ would be effectively equal to the initial (or physical) size of the crack $2a_o$. The load-displacement trace would be essentially linear up to the point at which the specimen fractures abruptly (see Fig. 4.5a). The plane strain fracture toughness K_{Ic} can be computed directly from the maximum load P_{max} or stress σ_{max} (*i.e.*, the load or stress at fracture) and the initial crack size a_o using Eqn. (4.7).

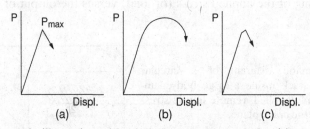

Figure 4.5. Schematic illustration of load-displacement records for (a) an overly large a_o (case I), (b) too small an a_o (case II), and (c) lower limit for an adequate a_o (case III).

CASE II. $2a_o < 1.5(K_{Ic}/\sigma_{ys})^2$ (a_o is too small).

It may be seen from Eqn. (4.9) that the applied stress σ would exceed the yield stress σ_{ys} before the applied stress intensity factor K_I reaches the fracture toughness K_{Ic}. In other words, the material is expected to yield before fracture. The specimen, therefore, would undergo gross plastic deformation before fracture, and the load-displacement curve would be obviously nonlinear (see Fig. 4.5b). Even though the specimen may fracture abruptly, with little or no prior crack extension, the stress field of the crack would not match that given by linear elasticity with an acceptable degree of accuracy. In this case, K_I calculated formally from Eqn. (4.5) using the maximum load cannot be regarded as a valid measure of the plane strain fracture toughness K_{Ic} of the material.

CASE III. $2a_o = A(K_{Ic}/\sigma_{ys})^2$

It is clear, so far, that the crack diameter is the characteristic dimension of the simple specimen (with a penny-shaped crack in an infinitely large body) under discussion. Based on cases I and II, there should be a useful lower limit for $2a_o = A(K_{Ic}/\sigma_{ys})^2$, where $A > 1.5$. This useful lower limit cannot be deduced theoretically at present because of the lack of a detailed understanding of the processes of fracture and the inability to model the deformation of real materials. It must be established experimentally through large numbers of K_{Ic} tests, covering a representative range of materials.

In this case, the load-displacement record may be somewhat nonlinear near the maximum load point, *i.e.*, near the fracture load (see Fig. 4.5c). Most valid tests of practical test specimens exhibit this behavior. The nonlinearity represents plastic deformation around the crack border, and slight (irregular) crack extension during the last stage of the test. If the extent of the nonlinearity is *not excessive*, then it can be ignored and K_{Ic} can be calculated from the maximum (or fracture) load and the initial crack diameter $2a_o$.

The question now is how much nonlinearity is considered to be not excessive. Formally, the nonlinearity should not exceed that which would correspond to an increase in the initial (or physical) crack diameter ($2a_o$) by the plane strain plastic zone correction factor; *i.e.*, by $2r_{Iy}$ (see Eqn. (3.49)). Physically, it is acceptance of the fact that a plastically deformed zone would develop at the crack tip, and its presence is equivalent to a change in the effective crack length at the onset of fracture from $2a_o$ to $2a_o + 2r_{Iy}$; *i.e.*,

$$2a_o \rightarrow 2a_o + 2r_{Iy} = 2a_o + \frac{1}{3\pi}\left(\frac{K_{Ic}}{\sigma_{ys}}\right)^2 \approx 2a_o + 0.1\left(\frac{K_{Ic}}{\sigma_{ys}}\right)^2 \qquad (4.10)$$

This stipulation on the allowable extent of plastic deformation at the crack tip cannot be used conveniently in fracture testing. The extent of deformation that would be allowed, however, is equivalent to a specification on the change in load-displacement curve at the maximum load point in relation to the initial slope (*i.e.*, from elastic to elastic-plastic deformation). This change in slope can be readily measured and is used in plane strain fracture toughness testing.

4.3.2 Practical Specimens and the "Pop-in" Concept

The aforementioned specimen with a penny-shaped crack and similar specimens facilitate simple and straightforward measurements of K_{Ic}, at least in principle. They are impractical, however, for a number of good reasons. These necessarily large specimens are inefficient with respect to the amount of material and the loading capacity of the testing machine that would be required. They may not reflect the actual microstructure and property of the size of material of interest, and cannot discern directional properties of the materials.

A variety of specimens, with a through-thickness crack, have been developed for measurement of the K_{Ic} of materials in different product forms (*e.g.*, bars, forgings, pipes, and plates). These specimens and their testing protocol are described in Test Method E-399 for Plane Strain Fracture Toughness of the ASTM [2]. They are more efficient and appropriate for the specific product form, but are conceptually and analytically more complicated. The complexities arise, first, because the specimen dimensions in relation to the crack are not large enough, the influence of specimen boundaries on the stress field of the crack can no longer be neglected. As such, the stress intensity factor (K_I) expressions that incorporate these boundary influences would be more complicated. Second, their most efficient use involves the utilization of specimens of nearly marginal thickness in which the fracture load may exceed that corresponding to K_{Ic}. The measure, therefore, depends on the proper exploitation of the so-called pop-in phenomenon at the onset of crack growth in these specimens; *i.e.*, when K_I reaches K_{Ic}.

The "pop-in" concept was first developed by Boyle, Sullivan, and Krafft [3] and forms the basis of the current K_{Ic} test method that is embodied in ASTM Test Method E-399 [2]. The basic concept is based on having material of sufficient thickness so that the developing plane stress plastic zone at the surface would not "relieve" the plane strain constraint in the midthickness region of the crack front at the onset of crack growth (see Fig. 4.2). It flowed logically from the case of the penny-shaped crack, as shown in Fig. 4.4 in the previous subsection.

A specimen of finite thickness may be viewed simply as a slice taken from the penny-shaped crack specimen (Fig. 4.6). As a penny-shaped crack embedded in a large body, the crack-tip stress field is not affected by the external boundary surfaces and plane strain conditions that prevail along the entire crack front. As a slice, however, the crack in this alternate specimen is now in contact with two free

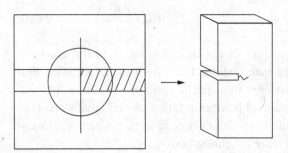

Figure 4.6. A finite-thickness specimen sliced from a large body that contains a penny-shaped crack.

surfaces. At the surface the state of stress at the crack tip is that of plane stress since no external traction is applied to these free surfaces. Because of the strain gradient associated with the crack, the elastic material ahead of the crack tip would exert constraint on lateral displacements along the crack front and thereby promote plane strain conditions in the interior region of the specimen. The effectiveness of this through-thickness constraint is reduced by the evolution of plastic deformation at the crack tip, particularly the development of plane stress plastic zones near the specimen surfaces.

If the specimen is very thick (*i.e.*, with thickness B much greater than the plastic zone size, or $B \gg (K_{Ic}/\sigma_{ys})^2$), the constraint condition along the crack front in the midthickness region is that of plane strain and is barely affected by plastic deformation near the surfaces. Abrupt fracture (*i.e.*, crack growth) will occur when the crack-tip stress intensity factor reaches the plane strain fracture toughness K_{Ic}. The load-displacement record, similar to that of the penny-shaped crack, is depicted by Fig. 4.7a.

For a very thin specimen (*i.e.*, with $B \ll (K_{Ic}/\sigma_{ys})^2$), the influence of plastic deformation at the surfaces will relieve crack-tip constraint through the entire thickness of the specimen before K_I reaches K_{Ic}. As such, the opening mode of fracture is suppressed in favor of local deformation and a tearing mode of fracture. The behavior is reflected in the load-displacement record by a gradual change in slope and final fracture, which could still be abrupt (see Fig. 4.7b), but the conditions of plane strain would not be achieved.

At some intermediate thickness, the relief of constraint is incomplete and the crack in the midthickness region can "jump forward" when K_I reaches K_{Ic}. This burst of growth is arrested because of plastic deformation along the near-surface portions of the crack and momentary unloading of the specimen caused by the change in specimen compliance with crack extension. This burst of crack growth, or crack "pop-in," produces a stepwise change in displacement in the load-displacement record (see Fig. 4.7c) and serves as the measurement point for K_{Ic}. The extent of load increase and further crack growth before final specimen fracture would depend on the thickness, the crack size, and the planar dimensions of the specimen.

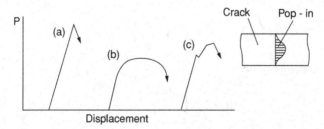

Figure 4.7. Schematic illustration of typical load-displacement records from finite-thick specimens used in plane strain fracture toughness measurement by using the "pop-in" concept (a) a very thick specimen, (b) a very thin specimen, and (c) a specimen with optimum thickness (Boyle, Sullivan, and Krafft [3]).

Clearly, there would be a minimum thickness to ensure the onset of (or momentary) plane strain crack growth at the midthickness region of a specimen, *i.e.*, the occurrence of pop-in. Because the relief of constraint is associated with plastic deformation near the specimen surface, the thickness requirement is expected to be a function of the plastic zone size and must be established experimentally; *i.e.*,

$$B \geq A_1 \left(\frac{K_{Ic}}{\sigma_{ys}} \right)^2 \tag{4.11}$$

This requirement specifically addresses the condition of "plane strain" over the midthickness region of the specimen at the onset of crack growth "instability." It complements those for crack size and planar dimensions of the specimen that ensure the applicability of (or the validity of using) linear fracture mechanics as an approximation. The difference in the basis for these requirements should be clearly understood.

The actual size requirements needed to be established experimentally, and were bounded by the work, principally on very-high-strength steels, through a special committee of the American Society of Testing of Materials (now Committee E-08 on Fracture and Fatigue, of the American Society of Testing and Materials). The supporting data, then reflecting interest in very-high-strength steels for aerospace applications, are published in an ASTM Special Technical Publication (STP 410), and are summarized here (see Figs. 4.8–4.10) [1]. Figure 4.8 (a–c) shows the influence of crack size, indicating the presence of a lower limit that is influenced by the material yield strength and fracture toughness. Figure 4.9 (a–c) shows the existence of a lower bound with respect to specimen thickness, which again depends on yield strength and fracture toughness. Figure 4.10 suggests that crack length does not represent a significant constraint, except in relation to stress gradients and stress levels in the uncracked ligament.

4.3.3 Summary of Specimen Size Requirement

In summary, the specimen size requirements for plane strain fracture toughness measurements are as follows:

Plane strain (pop-in) requirement

$$B \geq 2.5 \left(\frac{K_{Ic}}{\sigma_{ys}} \right)^2 \tag{4.12}$$

Elastic analysis requirement

$$a \geq 2.5 \left(\frac{K_{Ic}}{\sigma_{ys}} \right)^2 \tag{4.13}$$

$$W = 2a \geq 5.0 \left(\frac{K_{Ic}}{\sigma_{ys}} \right)^2$$

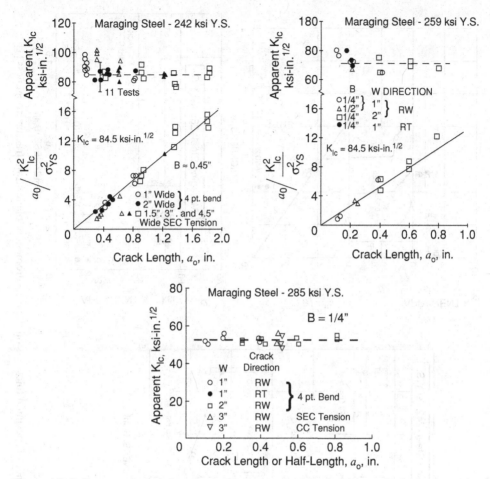

Figure 4.8. Influence of crack length on the measurement of plane strain fracture toughness [1].

The values were determined through experimentation. The fact that they are the same for "plane strain" and "elastic analysis" is coincidental. The specimen width requirement was based on an evaluation of the influence of the size of remaining ligament (*i.e.*, the uncracked portion of the cross section) and on the variation in K_I with crack length. The experimental results showed little dependence on ligament size. As such, the coefficient of 5.0 (or $a/W = 0.5$) was chosen to ensure good accuracy in the solution for K_I in relation to the precision in measuring crack length.

4.3.4 Interpretation of Data for Plane Strain Fracture Toughness Testing

This is the most demanding part of the measurement for fracture toughness. It is recommended that the readers familiarize themselves with the discussions in ASTM STP 410 [1], which captures the historical development of the methodology, and

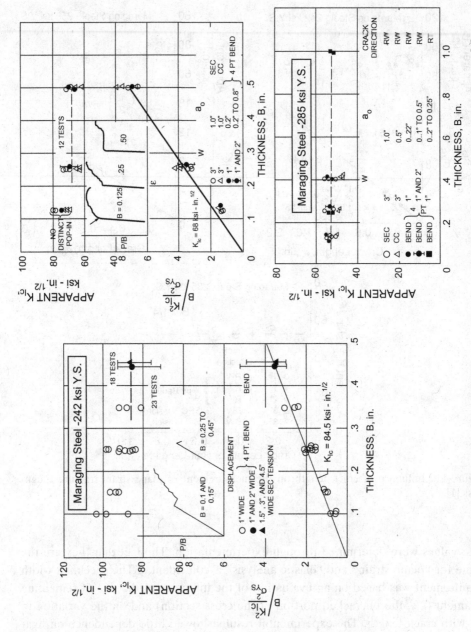

Figure 4.9. Influence of specimen thickness on the measurement of plane strain fracture toughness [1].

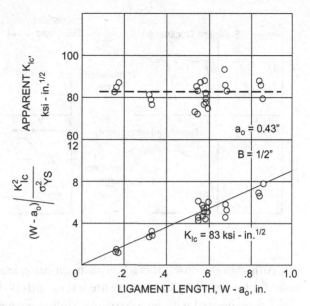

Figure 4.10. Influence of ligament length on the measurement of plane strain fracture
toughness [1].

ASTM Method E-399 and the supporting documents that document the evolution
of the method.

The onset of crack growth, and of fracture, is determined from an autographic
recording of the applied load and the crack-opening displacement; typical traces
are shown in Fig. 4.11. A displacement transducer, based on a strain-gage bridge
or a LVDT (linearly variable differential transformer) typically is used. Typical
load-displacement traces, reflecting the cracking response of test specimens, fall into
three types and are also shown in Fig. 4.11. Types a and b represent specimens that
meet dimensional requirements and are deemed to reflect valid tests of plane strain
fracture toughness K_{Ic}. Type c behavior represents a specimen that is too thin and,
therefore, would not yield valid K_{Ic}.

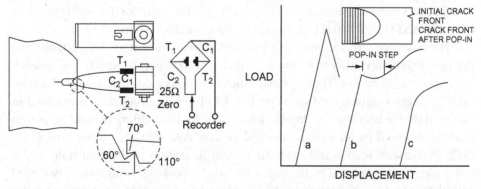

Figure 4.11. Typical strain gage-based crack-opening displacement gage (left), and typical
load-displacement traces observed during fracture toughness testing (right) [1].

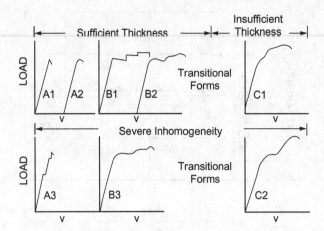

Figure 4.12. Representative load-displacement traces observed during fracture toughness testing that show "normal" behavior and those that reflect severe microstructural inhomogeneity or environmental effect [1].

Although the onset of crack growth, on rising load, serves as the measurement point for fracture toughness, there are test records that exhibit behavior that reflects artifacts associated with microstructural inhomogeneity and severe environmental sensitivity, respectively. They serve as "false" (low) indicators of fracture toughness, and are deemed to be invalid. Typical examples are shown in Fig. 4.12.

For the valid plane strain fracture toughness tests (Fig. 4.11) represented by type a behavior, the specimen dimensions are more than adequate relative to the size requirements given in Section 4.3.3 and there is no need for special interpretation. The K_{Ic} can be simply calculated from the maximum load and the initial crack length by using an appropriate K expression for the specimen. Load-displacement behavior, represented by type c, corresponds to specimens with insufficient thickness, and would not yield a valid measure of K_{Ic}. (The test, however, may be valid for measurement of fracture toughness under nonplane strain, or plane stress conditions, and will be considered in the discussion on crack growth resistance or R curves later.) Type b behavior, on the other hand, is associated with specimens having dimensions that are nearly minimal, or optimal, and is fundamental to the standard methods of test, such as ASTM Test Method E-399 for Plane Strain Fracture Toughness. The basic rationale for the prescribed procedure for data analysis is discussed in ASTM STP 410 [1], Appendix I, and summarized here.

The basic concern for analyzing the load-displacement trace is in establishing the permissible deviation from linearity that precedes pop-in, and the sufficiency of the pop-in indication. These considerations are discussed with the aid of a typical load-displacement record shown in Fig. 4.13 [1]. Various quantities involved in the analysis are also shown. Pop-in is indicated by the load maximum, or pop-in load P_p, followed by an increase in displacement Δv_p with decreasing load. (This slight decrease in load corresponds to a slight relaxation in the load train of the test apparatus engendered by the slight increase in specimen compliance associated with the pop-in crack extension.) The displacement v_i is that associated with the starting crack size at P_p if the specimen remains fully elastic. The indicated additional displacement Δv_i at P_p is the combined result of several effects (*e.g.*, crack-tip

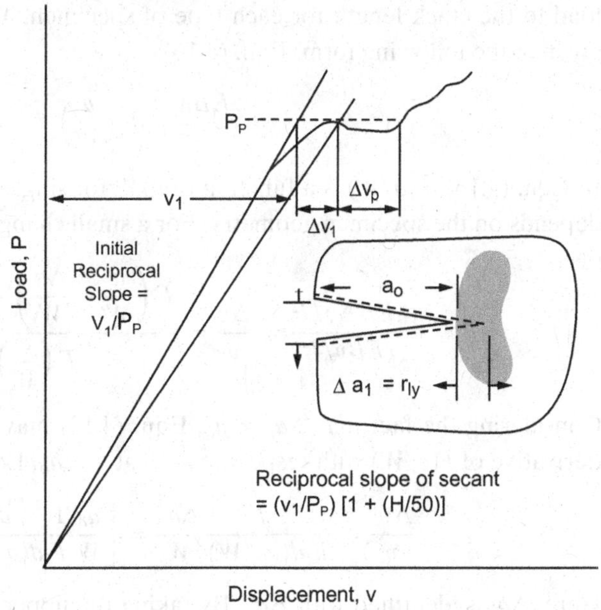

Figure 4.13. Typical load-displacement record showing quantities involved in the development of a data analysis procedure (after [1]).

plastic deformation, stable crack extension, etc.) and cannot be analyzed precisely. This deviation from linearity will be treated, instead, as an "effective" increment of crack extension Δa_i; in essence, the formation of the plane strain plastic zone at the crack tip.

For experimental purposes, the procedure must be cast in terms of measurable quantities from the load-displacement records (namely, the test data). As such, the physical quantities, such as crack increments, must be related to the changes in displacements, or in the slopes of the load-displacement records. To establish a permissible limit for $\Delta a_i / a_o$, it is assumed that Δa_i should not exceed the formally computed plane strain plastic zone correction term, namely

$$\Delta a_i \leq r_{Iy} \cong 0.05 \left(\frac{K_{Ic}}{\sigma_{ys}} \right)^2 \tag{4.14}$$

Also for valid tests, it is assumed (from Eqn. (4.13)) that

$$a_o \geq 2.5 \left(\frac{K_{Ic}}{\sigma_{ys}} \right)^2$$

Hence, for an acceptable test $\Delta a_i / a_o$ is given by Eqn. (4.15).

$$\frac{\Delta a_i}{a_o} \leq \frac{1}{50} \tag{4.15}$$

This condition may be expressed in terms of the displacement through the use of experimentally determined calibration curves that relate the displacement per unit load to the crack length for each type of specimen. The calibration relationship is given in the following form, Eqn. (4.16).

$$\frac{EBv}{P} = F\left(\frac{a}{W}\right) \tag{4.16}$$

In Eqn. (4.16), $F(a/W)$ is a function of a/W for single-edge-cracked specimens and depends on the specimen geometry. For a small change in Δv at constant load,

$$\frac{EB\Delta v/P}{EBv/P} = \frac{\Delta v}{v} = \frac{F\left(\dfrac{a}{W} + \dfrac{\Delta a}{W}\right) - F\left(\dfrac{a}{W}\right)}{F\left(\dfrac{a}{W}\right)} \tag{4.17}$$

Considering the fact that $\Delta a_i \ll a_o$, Eqn. (4.17) may be rewritten in terms of the derivative of $F(a/W)$ with respect to a/W at $a = a_o$, Eqn. (4.18).

$$\frac{\Delta v_i}{v} = \frac{1}{F}\frac{dF}{d(a_o/W)}\frac{\Delta a_o}{W} = \left[\frac{a_o}{W}\frac{1}{F}\frac{dF}{d(a_o/W)}\right]\frac{\Delta a_i}{a_o} \tag{4.18}$$

Here, Δa_i is identified with Δa_o. By taking the upper limit of $\Delta a_i/a_o = 1/50$ from Eqn. (4.15), the allowable limit of deviation from linearity in terms of displacements is

$$\frac{\Delta v_i}{v_i} \le \frac{1}{50}\left[\frac{a_o}{W}\frac{1}{F}\frac{dF}{d(a_o/W)}\right] = \frac{H}{50} \tag{4.19}$$

where H represents the quantity within the square brackets, and is derived from the experimentally derived calibration curve for each type of specimens. Values of H for selected specimens are shown in Fig. 4.14. It should be noted that the relationship between H and a/W will be independent of the gage length of the displacement gage, provided it is much smaller than the crack length.

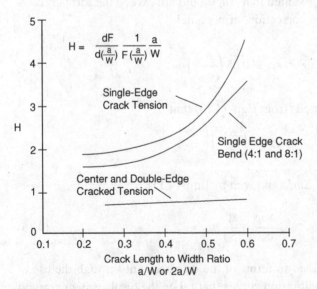

Figure 4.14. Calibration factors H for use in the analysis of load-displacement records in plane strain fracture toughness tests (after ASTM STP 410) [1].

For convenience in experimentation, the limitation on deviation from linearity may be expressed more conveniently in terms of the reciprocal slope of a secant line that connects the maximum load point P_p at pop-in and the origin. With reference to Eqn. (4.19)

$$\frac{v_i + \Delta v_i}{P_p} = \frac{v_i}{P_p}\left[1 + \frac{\Delta v_i}{v_i}\right] \leq \frac{v_i}{P_p}\left[1 + \frac{H}{50}\right] \tag{4.20}$$

For the recommended range of values of a_o/W of 0.45 to 0.55, a value for $H/50$ of 0.05 has been adopted for the testing of single-edge-cracked specimens. This is embodied in the so-called five percent slope offset method for establishing the "pop-in" load in ASTM Method E-399 for Plane Strain Fracture Toughness [2].

The question of how large a pop-in indication should be required can only be answered empirically. Ideally, the crack advance at pop-in should include sufficient material to be representative of the fracture property of the entire specimen. At this juncture, the consensus is that the increment of growth should be at least equal to the formally defined plane strain plastic zone correction factor (see Eqns. (4.14) to (4.16)). This limit may be related to the load-displacement record in a manner similar to that discussed previously.

In addition to these limitations, E-399 stipulates a maximum load level P_{max} beyond the pop-in load to ensure validity; namely, $P_{max}/P_p < 1.1$. It should be recognized that this requirement, along with those discussed earlier, presumes that the specimen dimensions could be tailored to the expected fracture toughness of the material to be evaluated. In reality, judgment must be used in the interpretation and utilization of data. Depending on applications, adjustments in specimen sizes may be essential to meet code requirements, and commitment to specimen design should only be made after preliminary evaluations.

Some commonly used specimens and stress intensity factor solutions are given in ASTM Method E-399 on fracture toughness testing [2]. Stress intensity factor solutions for other geometries are given in handbooks by Sih *et al.* [4] and Tada *et al.* [5], and may be determined numerically from commercially available finite-element codes (such as ANSYS [6]).

4.4 Crack Growth Resistance Curve

For applications involving materials in "thin" sections, the resistance to unstable crack growth is enhanced by the loss of through-thickness constraint and change in crack front contour, and the concomitant formation of shear lips, or plastic deformation, in the near-surface regions (see Fig. 4.15). Depending on the material thickness, the onset of crack growth on increasing loading may begin at $K = K_{Ic}$ and the crack plane essentially perpendicular to the loading axis, and gradually transition to partial or full-shear failure along forty-five degree inclined planes (commonly referred to as "shear lip formation") (see Figs. 4.15 and 4.16). The development of crack growth resistance is principally associated with this evolution of shearing

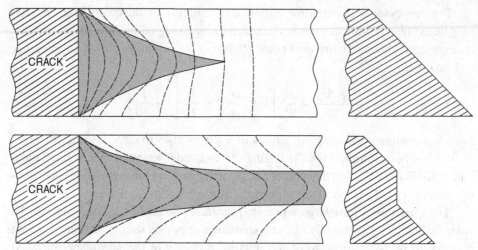

Figure 4.15. Schematic diagram showing the difference in the evolution of shear lips with crack growth that reflect the loss of through-thickness constraint through differences in material thickness, yield strengths, or fracture toughness [7].

mode of failure, or shear lip formation, and the increased work required for each increment of crack advancement.

With the onset of shear lip formation, it should be recognized that the problem is no longer one of mode I, or tensile-opening mode. For the sake of simplicity,

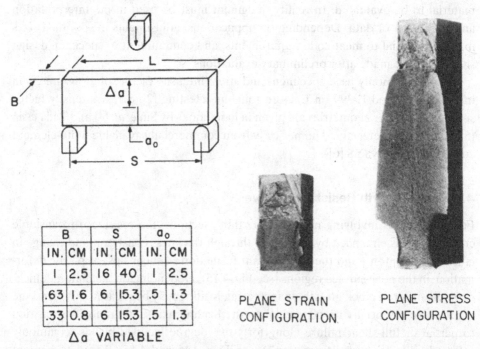

B		S		a_0	
IN.	CM	IN.	CM	IN.	CM
1	2.5	16	40	1	2.5
.63	1.6	6	15.3	.5	1.3
.33	0.8	6	15.3	.5	1.3

Δa VARIABLE

PLANE STRAIN CONFIGURATION

PLANE STRESS CONFIGURATION

Figure 4.16. Changes in fracture mode or extent of shear lip development with material thickness, with the thicker specimen (on the left) under conditions tending toward plane strain and the thinner specimen toward plane stress [7].

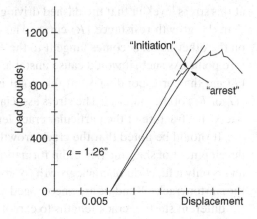

Figure 4.17. Typical load-displacement record showing stepwise evolution of crack growth resistance in a displacement-controlled rising load test [8].

and as a matter of engineering practice, however, the mode I framework has been "accepted" and adopted for engineering applications [2, 3]. The evolution of crack growth resistance and the concept of fracture toughness are developed within this framework. The values of crack growth resistance as a function of crack length (or crack extension) for a material, of a given thickness, are derived from the local peaks and valleys in an autographic recording of the applied load and the corresponding displacement across the crack (see Fig. 4.17). (The peaks correspond to the onset of (momentary) "unstable" crack extension, and the valleys correspond to "arrest" in crack growth engendered by the momentary "unloading" of the specimen from the crack growth-induced increase in its compliance associated with the "inertia" in the testing machine.

Figure 4.18 shows the crack growth resistance curve as a function of crack length (*i.e.*, the sum of the starting crack length and the individual crack growth increments) that had been constructed from the data (*i.e.*, the local peaks and valleys in the load-displacement trace) in Fig. 4.17. The increase in crack-driving forces (in terms of G) at two stress levels are shown as dashed line. At the lower stress level, the trend line shows the inadequacy of the driving force to continue crack growth

Figure 4.18. Crack growth resistance curve constructed from the load-displacement data in Fig. 4.14, showing the relationship between increasing stress, crack growth, and the evolution of crack growth resistance and fracture toughness [8].

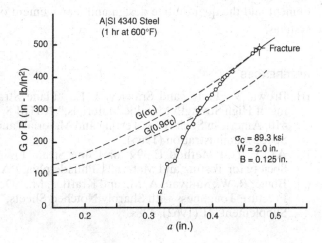

at this stress level, in that the dashed driving force curve crosses and becomes lower than the growth resistance (R) curve. The driving force at the higher stress level, on the other hand, becomes tangent to the R curve, and exceeds it beyond the fracture point. As such, it would cause unstable crack growth, or fracture. The value of G (and the corresponding K) at this point is identified with the fracture toughness (G_c or K_c) of the material. The stress associated with this failure point is the "critical stress" for fracture at the particular crack length.

It should be noted that the crack growth resistance curve is associated with the development of shearing (shear lip formation and growth) with crack prolongation, and is only a function of crack growth (or incremental increase in crack length from the starter crack). Readers are encouraged to construct G and R (or K and K_R) plots for different starting crack lengths to explore the influences initial crack size on the stress and crack size at fracture.

4.5 Other Modes/Mixed Mode Loading

In this chapter, the focus has been placed on the development and use of fracture mechanics in problems involving "homogeneous" materials under the tensile-opening mode (mode I) loading conditions. Specifically, the loading axis is, in principle, perpendicular to the crack plane and the crack growth direction. In fact, the methodology has been extended to include cases of fracture in "thin" sheets, which involves a fully shearing mode of failure, with the crack plane inclined at about forty-five degrees to the loading axis, but the crack growth direction perpendicular to it. By being consistent in the measurement of properties, and the use of the resulting data, this process has been used successfully for more than four decades.

A number of methods/theories have been developed over the years to treat this issue more rigorously (see, for example, Sih *et al.* [4]). The reality, however, is that these methods/theories are still limited, and cannot address cases that involve compression across the crack faces and the complexities associated with crack-face interactions. Here, therefore, the current methodology is being extended to the studies of material response, and to demonstrate its value in advancing materials development and the quantitative design and management of highly valued engineering systems.

REFERENCES

[1] Brown, W. F. Jr., and Srawley, J. E., "Plane Strain Crack Toughness Testing of High Strength Metallic Materials," ASTM Special Technical Publication 410, American Society for Testing and Materials and National Aeronautics and Space Administration (1965).

[2] ASTM Test Method E-399 for Plane Strain Fracture Toughness, American Society for Testing and Materials, Philadelphia, PA.

[3] Boyle, R. W., Sullivan, A. M., and Krafft, J. M., "Determination of Plane Strain Fracture Toughness with Sharply Notched Sheets," Welding Journal Research Supplement, 41 (1962), 428s.

[4] Sih, G. C., *et al.*, Stress Intensity Handbook, "Methods of Analysis and Solutions of Crack Problems," Mechanics of Fracture 1, G. C. Sih, ed., Noordhoff International Publishing, Leyden, The Netherlands (1973).

[5] Tada, H., Paris, P. C., and Irwin, G. R., "The Stress Analysis of Cracks Handbook," 3rd ed., ASME Press, New York (2000).

[6] ANSYS, Computer Code. Ansys, Inc., Canonburg, PA.

[7] Judy, R. W., Jr. and Goode, R. J., "Fracture Extension Resistance (R-Curve) Characteristics for Three High-Strength Steels, Fracture Toughness Evaluation by R-Curve Methods," ASTM Special Technical Publication 527, American Society for Testing and Materials, (1973) 48–61.

[8] Wei, R. P., unpublished results, Philadelphia, PA (1970).

5 Fracture Considerations for Design (Safety)

It is clear that two very different measures of "strength" in design must be considered: one measure deals with its resistance to inelastic deformation (yielding or "plastic flow") under stress and the other deals with fracture (at K_{Ic} or K_c). Each is viewed as an independent failure criterion. The specific requirement would depend on the design and configuration at locations where cracking might originate. In this chapter, the first effort at reconciling the criteria of design against yielding (inelastic deformation) and fracture, attributed to Irwin, is reviewed. Its impact on setting minimum fracture toughness requirements and the need for fracture mechanics-based design are discussed. A statistically based methodology for defining safety factors in design, proposed by J. T. Fong [1], is presented and is used in the rational definition and use of safety factors in design.

5.1 Design Considerations (Irwin's Leak-Before-Break Criterion)

As a conceptual experiment, one might consider a wide plate, with a central, through-thickness crack of length $2a$, under a uniformly applied tensile stress σ. One might further consider having both failure modes (fracture and yielding) occur at the same time. For fracture,

$$K_I = \sigma \sqrt{\pi a} \quad \text{or} \quad \sigma = \frac{K_I}{\sqrt{\pi a}}$$

where a is taken as the effective half-crack crack length, and includes the plastic zone correction factor. Assuming that yielding and fracture occur concurrently in plane strain, the estimated minimum fracture toughness would be:

$$(K_{Ic})_{\min} = \sigma_{ys} \sqrt{\pi a}$$

The minimum fracture toughness required would depend on the yield strength of the material, and the "expected" crack size, which is to be set by the efficacy and fidelity of regularly scheduled nondestructive inspections.

Figure 5.1. Schematic diagram showing Irwin's leak-before-break criterion.

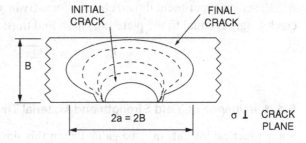

A conceptual illustration of a suggested alternative approach to this issue was given by G. R. Irwin for possible use in the design, material selection, and management of pressure vessels in power plants [2]. Here, because of safety considerations and environmental conditions, it would not be prudent to perform periodic inspections on the operating vessel. The "guiding criterion," here, became *a puddle of water on the plant floor (under the operating pressure vessel)*. The presence of this puddle is a clear indication that a crack had grown and penetrated the wall of the vessel. This approach requires only due diligence, no instruments and skilled inspectors. The safe-design criterion is then built on the requirement that the material's fracture toughness must be sufficient to preclude fracture at this, yet to be defined, crack size and stress level.

In Irwin's suggested approach, the crack in the pressure vessel is modeled as a central crack in an infinitely large plate, subjected to uniform traction. The length of the crack is aligned parallel to the axis of the vessel, with the crack plane perpendicular to the circumferential direction (*i.e.*, oriented perpendicular to the hoop stress). It is assumed that, at the onset of leakage, the overall length of the crack is equal to twice the vessel-wall thickness (*i.e.*, $2a = 2B$), Fig. 5.1, and the crack is treated as a through-thickness crack at this point. For "optimal" design, it is assumed that yielding and fracture will occur at the same time. In other words, the hoop stress (σ) and the crack-tip stress intensity factor (K) will reach the material's yield strength (σ_{ys}) and its minimum required fracture toughness $(K_c)_{min}$ concurrently. The required minimum fracture toughness is given in terms of the yield strength and effective crack size a as follows:

$$(K_c)_{min} = \sqrt{\pi a_{eff}} \quad \text{or} \quad a_{eff} = \frac{(K_c)^2_{min}}{\pi \sigma^2_{ys}} \tag{5.1}$$

The effective crack length, however, is the sum of the length of the physical crack and the plastic zone correction factor. As such,

$$a_{eff} = a + r_y = a + \frac{(K_c)^2_{min}}{2\pi \sigma^2_{ys}} \tag{5.2}$$

By combining Eqns. (5.1) and (5.2) and solving for a, the physical crack size (a) at the onset of leakage/fracture becomes:

$$a = \frac{(K_c)^2_{min}}{2\pi \sigma^2_{ys}} \quad \text{or} \quad (K_c)^2_{min} = 2\pi \sigma^2_{ys} a \quad \text{or} \quad (K_c)_{min} = \sigma_{ys}\sqrt{2\pi a} \tag{5.3}$$

Based on experimental/field observations, Irwin assumed that typically the half-crack length is equal to the plate thickness, and thus:

$$(K_c)_{min} = \sigma_{ys} \sqrt{2\pi\, B} \tag{5.4}$$

5.1.1 Influence of Yield Strength and Material Thickness

Some practical insights may be gained from this simple exercise. For example, the minimum fracture toughness required for material of different yield strengths and plate thicknesses may be estimated (see Table 5.1 below). It is evident that greater fracture toughness (or energy dissipation capability) would be required at the higher yield strengths and material thicknesses. This is consistent with the fact that greater amounts of elastic energy would have been stored in the structure under these conditions. The unfortunate fact is that energy dissipation capabilities are sacrificed for strength, and must be judiciously considered in design and material selection.

5.1.2 Effect of Material Orientation

Most metallic materials used in manufacturing and construction have undergone melting and casting, and subsequent metal-working processes, such as forging and rolling. They contain precipitates that are formed from the addition of alloying elements that increase their strengths, but they can also form other particles with other elements that can improve their machinability. These particles tend to segregate to grain boundaries, and tend to degrade the material's fracture resistance (*i.e.*, fracture toughness) in relation to its grain size and orientation. As such, their fracture toughness, along with their strengths, depends on their orientation. These orientations of sheet and plate products are typically defined in relation to their primary rolling (forming) direction, which is designated as the longitudinal (L) direction. The width direction is designated as the transverse (T), or long-transverse direction. The thickness direction is also transverse to the rolling direction, but is designated as the "short-transverse" (S) direction. These directions are indicated in Fig. 5.2.

Table 5.1. *Minimum fracture toughness requirement as a function of yield strength and thickness based on Irwin's "leak-before-break" criterion*

B (in.)	$(K_c)_{min}$ (ksi-in$^{1/2}$)			
	50 ksi	100 ksi	150 ksi	200 ksi
0.1	40	80	120	160
0.5	90	180	270	360
1.0	125	250	375	500
4.0	250	500	750	1000
8.0	355	710	1050	1420
12.0	435	870	1300	1740

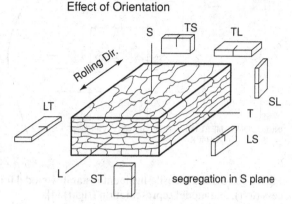

Figure 5.2. Schematic illustration of grain structure in a rolled plate, along with designations for the rolling or longitudinal (L), transverse (T), and short-transverse or thickness (S) directions, and the associated cracking planes and crack growth directions per ASTM Method E-399 [3].

The orientations of the crack plane and crack growth direction are defined in terms of these definitions; for example, LT designates a specimen that has its crack in the TS plane and is to be loaded in the L direction.

From the material properties perspective, fracture toughness can depend strongly on the orientation of the crack plane and the direction of crack growth. This orientation dependence is principally a result of heterogeneities introduced by the processes used in its creation. For example, in casting, inclusion particles formed during solidification tend to be entrapped between solidifying dendrites and also swept into the center of the ingot, which is the last region to solidify. During rolling, for example, these particles tend to be "broken" and distributed along boundaries that become parallel to the rolling, or the S plane. As such, the lowest fracture toughness tends to be associated with the short-transverse (SL and ST) orientations.

Because the primary stresses are typically applied in the plane of the plate (termed in-plane loading), the primary orientations of interest are the LT and TL orientations; *i.e.*, with loading in the longitudinal (L) direction and crack growth in the transverse (T) direction, for LT, or vice versa, for TL. Fracture toughness tends to be higher in the LT (*vis-à-vis* the TL) orientation to reflect the influences of rolling texture. For a surface crack that lies in the TS plane and grows in the through-thickness (S) direction, on the other hand, it may produce delamination perpendicular to the growing crack, and result in crack "deflection," or "blunting." This phenomenon can produce "false" indications of fracture toughness and must be treated with care. On the other hand, the phenomenon itself may be used for producing laminated structures that provide enhanced fracture toughness for thick-section applications.

5.2 Metallurgical Considerations (Krafft's Tensile Ligament Instability Model [4])

To provide a linkage between fracture toughness and some controlling microstructural parameter of the material, a simple model was proposed by Krafft [4]. The model envisioned the presence of an array of "inclusions" aligned ahead of the

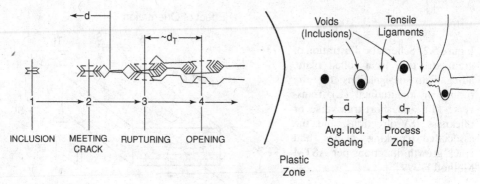

Figure 5.3. Krafft's tensile-ligament instability model for fracture. Schematic of physical process (left), and model representation (right) [4].

crack tip and weakly bonded to the metal matrix. Under load, the particles immediately ahead of the crack tip debond from the metal matrix, and allow cavities/voids to form and grow around these particles, thereby isolating "ligaments" of material that can grow under load (see Fig. 5.3). With increasing loads, these ligaments are strained until they reach the point of onset of tensile deformation instability (*i.e.*, corresponding to the maximum load point in a tensile test). Krafft identified this point with the onset of crack growth instability, and thereby established a relationship between the fracture and deformation properties of the material, and the relationship between the fracture toughness of a material with it mechanical properties and some pertinent microstructural character.

The essence of Krafft's model involves the relationship between the onset of crack growth with that of plastic flow instability in these tensile ligaments ahead of the crack tip. Assuming that the strains within the crack tip plastic zone are constrained by the surrounding elastic material, the strain inside the plastic zone would follow a $1/r^{1/2}$ singularity as dictated by the surrounding elastic stress-strain field; namely,

$$\varepsilon_{yy} = \frac{\sigma_{yy}}{E} = \frac{K_I}{E\sqrt{2\pi r}} \tag{5.5}$$

He proposed that crack growth instability, or fracture, would correspond to the onset of plastic flow instability (or necking) in the tensile ligament(s) at the crack tip; or when $\varepsilon \to \varepsilon_m$ at $r = d_T$. Thus,

$$K_{Ic} = E\varepsilon_m \sqrt{2\pi d_T} \tag{5.6}$$

The essential challenge is to identify the quantities ε_m and d_T in relation to the appropriate deformation properties and metallurgical characteristics of the material.

For this purpose, Krafft considered a simple tensile test, Fig. 5.4, where the "true stress" σ is given in terms of the applied load P and the "current" cross-sectional area A of the specimen; namely:

$$\sigma = \frac{P}{A} \text{ or } P = \sigma A \quad \text{and} \quad dP = \sigma dA + Ad\sigma \tag{5.7}$$

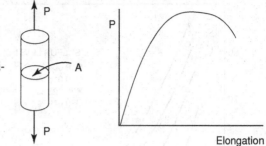

Figure 5.4. Schematic diagram of a uni-axial tensile test.

With increasing straining, P changes with σ and A in accordance with $dP = \sigma dA + A d\sigma$. At the maximum load point, the slope of the load-displacement curve would be zero, or $dP = 0$. Thus,

$$\frac{d\sigma}{\sigma} = -\frac{dA}{A} = 2v d\varepsilon$$

where v is the Poisson's ratio. For constant-volume plastic deformation (where $v = 1/2$) at the onset of plastic flow instability:

$$\frac{d\sigma}{\sigma} = 2v d\varepsilon = d\varepsilon \quad \text{or} \quad \frac{d\sigma}{d\varepsilon} = \sigma$$

For a power-hardening material that obeys the stress-strain relationship, $\sigma = k\varepsilon^n$, where n is the strain-hardening exponent, the slope of the stress-strain curve is given by Eqn. (5.8):

$$\frac{d\sigma}{d\varepsilon} = \frac{d}{d\varepsilon}(k\varepsilon^n) = nk\varepsilon^{n-1} = n\frac{k\varepsilon^n}{\varepsilon} = \frac{n\sigma}{\varepsilon} \tag{5.8}$$

Given that $d\sigma/d\varepsilon = \sigma$ and $\varepsilon = \varepsilon_m$ at the onset of deformation instability, then:

$$\frac{d\sigma}{d\varepsilon} = \frac{n\sigma}{\varepsilon_m} \equiv \sigma; \quad \text{or} \quad \frac{n}{\varepsilon_m} = 1, \quad \text{or} \quad \varepsilon_m = n$$

Hence,

$$K_{Ic} = En\sqrt{2\pi d_T} \quad \text{for "ductile" failure}$$
$$K_{Ic} = E\varepsilon_{\max}\sqrt{2\pi d_T} \quad \text{for "brittle" failure} \tag{5.9}$$
$$\text{where } \varepsilon_{\max} = \varepsilon_{tb} < n$$

The presumption here is that "brittle" failure would occur sooner, and at a ligament strain that is much lower than the "necking" strain in the material. This presumption is yet to be fully tested.

Physical support of this hypothesis came shortly after the publication of Krafft's model. Birkle et al. [5] were studying the influence of sulfur level (an impurity element present in steels) on the fracture toughness of very-high-strength

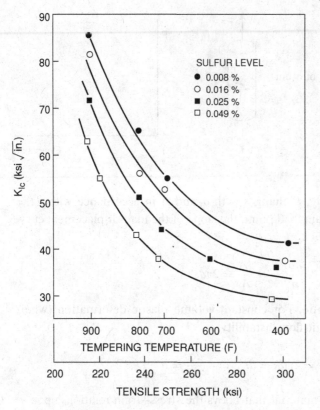

Figure 5.5. Influence of sulfur level on plane strain fracture toughness in a 0.45C-Cr-Ni-Mo steel at different tempering temperatures [5].

(0.45C-Cr-Ni-Mo) steels as a function of sulfur concentration. Their findings showed a systematic dependence of plane strain fracture toughness (K_{Ic}) on sulfur level at different tempering temperatures (see Fig. 5.5). This support came in the form of fractographic measurements of the average spacing of MnS (manganese sulfide) inclusion particles in these steels (*cf.* Fig. 5.6, and Table 5.2). Independent measurements of strain-hardening exponents showed little or no influence of sulfur content [5]. This correlation had fostered further extension into other areas, for example, creep-controlled crack growth [6–8]. Extension of this level of understanding into the fracture of more "brittle" alloys needs to be made, and requires further insight and collaboration among the mechanics and materials disciplines.

5.3 Safety Factors and Reliability Estimates

At this juncture, it is appropriate to re-examine the concept and usage of "*safety factors*" in design, particularly with respect to the quantitative justification of their underlying basis, or bases, and to the assumption (*presumption*) of the absence/presence of pre-existing crack-like "*damage.*" Traditional designs are based on the use of "*safety factors,*" applied against the material's yield strength or tensile strength to establish the maximum allowable stresses in design. Consensus standard safety factor(s), guided by specific industries or technical societies (*e.g.*, the American Society of Mechanical Engineers and the American Petroleum Institute)

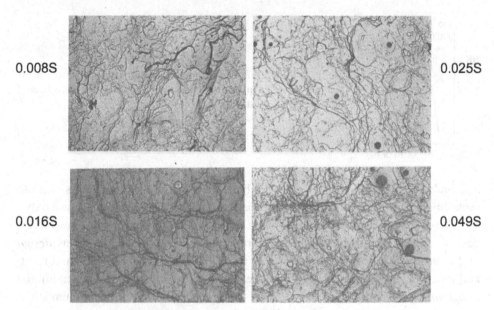

Figure 5.6. Fractographic evidence on 0.45C-Cr-Ni-Mo steel, tempered at 800°F, showing the difference in inclusion content at different sulfur levels [5].

are typically used. These "*safety factors*" are experientially based and are applied against minimum properties of given materials, such as their tensile strengths or yield strengths, and do not address the *risk* for failure in a probabilistic sense.

The planning and construction of large-diameter pipelines to transport crude oil from the Aleyeska oil fields in northwestern Alaska to the seaport at Anchorage raised significant concerns regarding the fracture safety and reliability of such systems, particularly with respect to the impact of such a failure in relation to the pristine (*e.g.*, Alaskan) environment. In the early 1980s, Dr. Jeffrey Fong of the National Bureau of Standards began to raise concerns on the need to quantify the design of engineered systems so that the reliability and safety of such systems could be quantified. These concerns were translated into a special symposium, under the auspices of the American Society for Testing and Materials (ASTM) in 1984, to raise awareness of the issues with the scientific and engineering communities. The proceedings were published in 1988 as ASTM Special Publication 924, "Basic Questions in Fatigue," Volumes I and II [1]. Here, a synopsis of the philosophical issue is summarized to highlight the fundamental issues. Readers are encouraged to keep these issues in

Table 5.2. *Correlation between process zone size and average inclusion spacing [5]*

Steel	S(w/o)	K_{Ic} MPa-m$^{1/2}$	d_T μm	$d_{av} = (A/N)^{1/2}$ μm
A	0.008	72	5.7	6.1
B	0.016	62	4.1	5.4
C	0.025	56	3.5	4.4
D	0.049	47	2.4	3.7

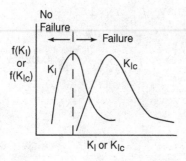

Figure 5.7. Challenges in design attributed to uncertainties in loading and material's property (as reflected by K_I and K_{Ic}).

mind as they continue to explore the subject area of fracture mechanics and life cycle engineering. For this purpose, the discussion here will focus on fracture only.

The essential challenge is to provide a rational definition of safety factor for design, and quantitative estimations of reliability. Fong coined the terms *design under uncertainty* and *design under risk* to distinguish between the then current, and much of the current, design practices, and *design under risk*. For traditional *design under uncertainty*, a safety factor is used to *ensure* that the maximum stress on a component, or a system, made from some material, does not exceed the controlling minimum property of that material. In such designs, the risk for failure is not, and could not be quantified because of lack of information. For *design under risk*, estimations of risk for failure, along with estimations of the confidence level these estimates, are to be provided. In the fracture mechanics context, this challenging problem involves dealing with uncertainties in, for instance, load level, crack size distribution (as characterized by, *e.g.*, K_I), and uncertainties in material properties (as characterized by, *e.g.*, K_{Ic}), as depicted schematically as probability density functions in Fig. 5.7.

The development of methodologies for design and regulation is depicted pictorially in terms of the building of a multispan bridge across a river, or a ravine, in Fig. 5.8. The development and formulation of design methodology is to be supported

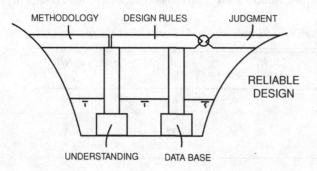

Figure 5.8. Conceptual bases and processes involved in the development of reliable design.

by the quantitative understanding of materials and their interactions of loading and environmental influences. The formulation of design methodology and rules is to be based on quality data bases on material-loading-environmental influences. The arrival at final "acceptable and reliable designs" is made based on regulatory judgments by the responsible rule-making bodies, and is not considered here. In the following discussion, the choice of distribution function for material property and the sample size on which the design is based are considered.

To highlight the need and the impact of the statistically based considerations, *safety factor (S.F.)* is now defined on the basis of the *mean* and some selected *minimum* property (*e.g.*, allowable design strength).

$$S.F. = \frac{\text{mean}}{\text{minimum}} = \frac{\overline{X}}{X^*} \tag{5.10}$$

X^* is associated with some probability for failure, which depends on the choice of the distribution function. For illustration, (a) the influence in choice of the distribution function is considered in terms of the *Normal (or Gaussian)* and the *Weibull* distributions, and (b) the influence of sample size is examined using the *Normal* distribution.

5.3.1 Comparison of Distribution Functions

The Normal (or Gaussian) and Weibull distribution functions are most commonly used in engineering design. The normal distribution is usually expressed in one of the following two functional forms:

Normal (Gaussian) Distribution

$$f(x) = \frac{1}{\sqrt{2\pi}\sigma} \exp\left[-\frac{(x-\mu)^2}{2\sigma^2}\right] \quad \text{Probability density}$$

$$F(x) = \int_x^\infty f(x)dx \quad \text{Probability of survival} \tag{5.11}$$

where μ is the mean and σ is the standard deviation of the distribution. In practice, μ and σ are replaced by the average and standard error, $\bar{x}$ and s, respectively. For comparison, the Weibull distribution function is written in one of two (differential and integral) forms:

Weibull Distribution

$$f(x) = \frac{b}{x_a - x_o}\left[\frac{x-x_o}{x_a-x_o}\right]^{b-1} \exp\left[-\left(\frac{x-x_o}{x_a-x_o}\right)^b\right]; \quad \text{for } x \geq x_o$$

$$f(x) = 0; \quad \text{for } x < 0 \tag{5.12}$$

$$F(x) = \exp\left[-\left(\frac{x-x_o}{x_a-x_o}\right)^b\right]; \quad x \geq x_o \text{ Probability of survival}$$

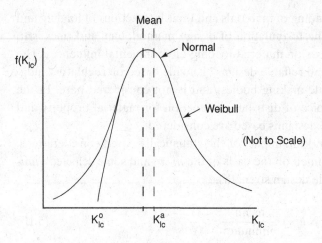

Figure 5.9. Schematic illustration on the choice of distributions in the design and management of engineered systems.

where x_o is the *minimum value* (giving explicit recognition to the existence of a minimum value to the quantity of interest), x_a is the *characteristic value* (corresponding to the 63.2 percentile point of the cumulative distribution), and b is the *shape parameter*. Through the use of these two "commonly" used distribution functions, the philosophical and practical impact of their choice on the design and management of engineered systems may be explored. Here, the impact of choosing Gaussian and Weibull distributions is first considered, followed by an examination of the choice of sample size on design through the use of the Gaussian distribution.

Figure 5.9 is a schematic comparison between the Normal (Gaussian) and Weibull distributions for plane strain fracture toughness K_{Ic} of a material in terms of the respective probability density distributions, $f(K_{Ic})$. The minimum and characteristic values of the Weibull distribution are designated as K_{Ic}^o and K_{Ic}^a, respectively. For this comparison, the Normal distribution is "known" (*i.e.*, not estimated) and is characterized by its mean (μ) and standard deviation (σ). For $b = 3.6$ in the Weibull distribution, Fig. 5.9 shows that the two distributions generally conform to each other, except at the low end. At an applied stress intensity factor K_I that is equal to or less than K_{Ic}^o, the Weibull distribution would lead to a prediction of zero probability for failure, whereas the use of the Gaussian distribution would predict some finite probability for failure, no matter how small. Based on broad/historical engineering and design experience, the existence of certain minimum properties is accepted. As such, distribution functions, such as the Weibull distributions, are commonly used. The specific choice of a distribution is determined by industry and regulatory bodies and is beyond the scope of this discussion.

5.3.2 Influence of Sample Size

Once the distribution function for the design property has been selected, the next step is to establish a design allowable. Such a value is based on the determination/estimation of some specific "design" property and its variability. For reliability analyses, the requisite property, or properties, must be quantified and statistically analyzed and documented, within a lot and over the lots of material that would be utilized over the product's life cycle. Determination of a design allowable would also involve cost and regulatory concerns, and the choice of an appropriate distribution function, which are beyond the scope of this discourse. Here, the process and its implication are considered through the selection of a "safety factor" for design. The selection of the "safety factor" is based on the design of a product that is "safe" within the context of prudent/design usage, and involves the choice of a distribution function. The true mean value and variance of the distribution function, indeed the function itself, are not (usually) known *a priori*, and must be estimated from experimental measurements. This estimation is fairly complex and can introduce uncertainties. Here, for simplicity, the philosophical framework and consequences are examined through the use of the Normal (Gaussian) distribution, as an example, in estimating the effect of sample size on design safety factor.

Traditionally, safety factor is defined as a "knock down" factor to reduce the allowable design stress to a lower level than the cogent property (say, the minimum tensile strength) of the material. Recognizing the fact that such a property can vary from point to point within a product form, and from lot to lot, such a property is grouped and represented through an appropriate statistical distribution, and characterized through the appropriate statistical parameters. If the population mean and variance (μ and σ) are known, the lower bound value of the property X (namely, $\overline{X}$) can be defined as $X^* = \mu - \beta\sigma$, the value of β is chosen to provide a "measure" that the probability that X will be below (or above) X^*. A safety factor ($S.F.$) may then be defined as:

$$S.F. = \frac{\overline{X}}{X^*} = \frac{\mu}{\mu - \beta\sigma}$$

For example, for $\beta = 3.0$, the $S.F.$ would provide a probability of failure of 1.35×10^{-3} or 0.135%, or a 99.86% probability of survival, which are "well defined" (assuming, of course, that the "property" is appropriately represented by the distribution). But, because the mean μ and standard deviation σ are both not known, safety factors must be defined in terms of their estimates based on n measurements (*i.e.*, on $\overline{X}_{(n)}$ and $S_{(n)}$), with some defined confidence level for the estimates, where both the population mean (μ) and the standard deviation (σ) would be estimated from the "limited" measured data.

As a numerical illustration, it will be assumed that the standard deviation σ of the Normal distribution is known, and the mean μ of the distribution is to be estimated from the sample average, say $\overline{X}_{(n)}$, from n measurements. The sample average $\overline{X}_{(n)}$ is a random variable with a standard deviation equal to $\sigma/\sqrt{n}$. The mean (μ) of the distribution is estimated by Eqn. 5.13 and is given by:

$$\mu = \overline{X}_{(n)} \pm \frac{\alpha\sigma}{\sqrt{n}} \tag{5.13}$$

where the confidence level of the estimate is determined from the Normal distribution table through the choice of α. For a confidence level of 99 percent, for example, $\alpha = 2.33$.

Table 5.3. *Effect of sample size*

n	$\sigma = 2.5$	$\sigma = 5.0$	Weight penalty
5	1.25*	1.68*	34%
10	1.23(2%)**	1.60(5%)**	30%
100	1.19(5%)**	1.48(12%)**	24%

* Not quite realistic because of small sample size.
** Improvement over $n = 5$.

Note that, in practice, the mean and variance of the distribution (μ and σ) are not precisely known, and are estimated from n measurements. As such the lower-bound estimate of the mean, at a prescribed confidence level, used to represent the mean of the distribution, and some multiple of the standard deviation (or standard error) is used to define the design allowable, namely,

$$\mu_{lbe} = \overline{X}_{(n)} - \frac{\alpha\sigma}{\sqrt{n}} \text{ at confidence level defined by } \alpha \qquad (5.14)$$

An equivalent safety factor, based on the measurement sample size n and a probability of failure determined by $\beta\sigma$, is then:

$$S.F._{\cdot(n)} = \frac{\overline{X}_{(n)}}{\overline{X}_{(n)} - \frac{\alpha\sigma}{\sqrt{n}} - \beta\sigma} \text{ vs. } \frac{\mu}{\mu - \sigma} \qquad (5.15)$$

with the probability of failure determined by $\beta\sigma$. The safety factor, utilized here, is used to account for variability in material properties, and does not account for design- and utilization-related factors.

The influences of sample size and material quality (or property variability) are illustrated in Table 5.3. The property is identified with plane strain fracture toughness (*i.e.*, $\overline{X}_{(n)} = \overline{K}_{Ic(n)} = 50$ ksi $\sqrt{\text{in.}}$, with $\sigma = 2.5$ or 5.0 ksi $\sqrt{\text{in.}}$ to reflect two different levels of variability, or manufacturing control). The results show marginal improvements with increasing number of tests to characterize variability in the property data, and substantial improvement in variability with quality control. It is recognized that the values shown at the lower numbers of test samples (say, $n = 5$) are inappropriate, but they do convey the need for a quantitative basis for design and reliability analyses.

5.4 Closure

In the foregoing chapters, the development and utilization of linear fracture mechanics in the design and management of engineered systems, from the fracture safety perspective, have been summarized. The subject matter represents an expansion from traditional mechanical design, and recognizes the need to treat the integrity and safety of structures and systems that contain cracks or cracklike inhomogeneities, particularly with respect to sudden fracture. Important as the subject is, the majority of the engineering problems involve progressive damage nucleation, growth, and accumulation. Solutions to these problems require a comprehensive understanding of the complex interactions between mechanical, thermal, and chemical forces with the material to effectively develop, utilize, and manage materials under the combined influences of these forces.

REFERENCES

[1] Fong, J. T., in *Basic Questions in Fatigue*, ASTM Special Publication 924, Volumes I and II. American Society for Testing and Materials, Philadelphia, PA (1988).

[2] Irwin, G. R., "Linear Fracture Mechanics, Fracture Transition and Fracture Control," Journal of Engineering Fracture Mechanics, 1 (1968), 241–257.

[3] ASTM Test Method E-399, ASTM Method of Test for Plane Strain Fracture Toughness, American Society for Testing and Materials, Philadelphia, PA.

[4] Krafft, J. M., "Correlation of Plane Strain Crack Toughness with Strain Hardening Characteristics of a Low, a Medium, and a High Strength Steel," Applied Materials Research (1964), 88.

[5] Birkle, A. J., Wei, R. P., and Pellissier, G. E., "Analysis of Plane-Strain Fracture in a Series of 0.45C-Ni-Cr-Mo Steels with Different Sulfur Contents," Trans. Quarterly of ASM, 59 (1966), 981.

[6] Landes, J. D., and Wei, R. P., "Kinetics of Subcritical Crack Growth and Deformation in a High Strength Steel," J. Eng'g Materials and Technology, Trans. ASME, Ser. H, 95 (1973), 2.

[7] Yin, H., and Wei, R. P., "Deformation and Subcritical Crack Growth under Static Loding," Materials Science and Engineering, A119 (1989), 51–58.

[8] Wei, R. P., Masser, D., Liu, H. W., and. Harlow, D. G., "Probabilistic Considerations of Creep Crack Growth," Materials Science and Engineering, A189, (1994), 69–76.

6 Subcritical Crack Growth: Creep-Controlled Crack Growth

6.1 Overview

In the foregoing chapters (2 through 5), the essential framework for linear elastic fracture mechanics is introduced. Within this framework, the presence (or preexistence) of a crack, or crack-like damage is assumed, and the driving force for its growth is given by an appropriate stress intensity factor (K_I), or strain energy release rate (G_I), that reflects the size, shape, and location of the damage relative to the loading. These chapters address, however, only the first of the "customer's questions" raised in Chapter 1; namely, "How much load will it carry?" They serve only as a basis for the design and management of engineered systems to guard against catastrophic failure.

Customer's Questions
- How much load will it carry, with or without cracks? (structural integrity and safety)
- How long will it last, with and without cracks? (durability)
- Are you sure? (reliability)
- How sure? (confidence level)

At loadings that are below that required for fracture, the next question is whether the damage can grow through time-dependent (subcritical crack growth) processes that lead to the progressive loss of design strength and reliability, and increase the chances for failure. The modes of subcritical crack growth in inert and deleterious environments are shown in Table 6.1. Subcritical crack growth under statically applied loads in deleterious environments (stress corrosion cracking), and fatigue crack growth under cyclically applied loads (in benign and deleterious environments), or fatigue and corrosion fatigue, were and are readily accepted. The possibility that creep-controlled crack growth can occur at or near room temperature was not universally recognized and accepted by the engineering/materials community in the mid-1960s, but is now well accepted.

In terms of life prediction and sustainment planning, it is essential to incorporate these processes for progressive damage (see Table 6.1). The first process in each

Table 6.1. *Categories of subcritical crack growth*

Loading condition	Inert environment	Deleterious environment
Static or sustained	Creep crack growth (or internal embrittlement)	Stress corrosion cracking
Cyclic or varying	Mechanical fatigue	Corrosion fatigue

group (namely, creep crack growth and mechanical fatigue) is conducted in an inert environment, and the remainder (namely, stress corrosion cracking and corrosion fatigue) in deleterious environments. The service life of a structure, or component, is determined by the growth of a crack-like damage from some initial size to a critical size to cause fracture. The overall process is to be incorporated into some form of "life prediction and sustainment planning" (see Fig. 6.1), where the damage function $D(x_i, y_i, t)$ (or crack) is a function of the internal (x_i) and external (y_i) variables. In this chapter, creep-controlled crack growth is considered. The impact of fatigue, corrosion fatigue, and stress corrosion cracking and crack growth at elevated temperatures are discussed separately in subsequent chapters.

6.2 Creep-Controlled Crack Growth: Experimental Support

In the mid-1960s, Li, *et al.* [1] documented the occurrence of subcritical crack growth under sustained load in high-purity, dry argon in high-strength steels. They showed that crack growth exhibited transient, steady-state, and tertiary crack growth, akin to creep deformation, and suggested that crack growth under sustained loading may

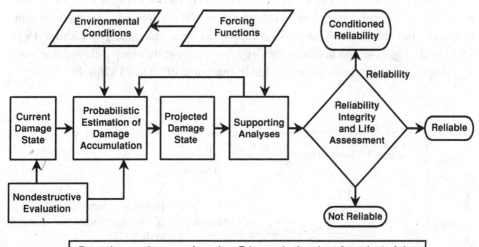

Figure 6.1. Schematic diagram depicting the essential processes for life prediction and sustainment planning [1].

be controlled by deformation processes (creep) occurring in the crack-tip process zone. Because crack growth occurred at K_I levels well below the fracture toughness of the material, experts in the then fracture mechanics and materials community could not agree whether such growth could indeed occur. As a result, this paper was withdrawn and was never published.

In a subsequent series of experiments, Landes and Wei [2] demonstrated that the phenomenon is real, and modeled the crack growth response in terms of creep deformation rate within the crack-tip process zone. The effort has been further substantiated by the work of Yin *et al.* [3]. The results and model development from these studies are briefly summarized, and extension to probabilistic considerations is reviewed. It is hoped that this effort will be extended to understand the behavior of other systems, and affirm a mechanistic basis for understanding and design against creep-dominated failures. The author relies principally on the earlier works of Li *et al.* [1], Landes and Wei [2], Yin *et al.* [3], Krafft [4] and Krafft and Mulherin [5]. The findings rely principally on the laborious experimental measurements by Landes and Wei [2], and the conceptual modeling framework by Kraftt [4]. Here, the Landes and Wei formulation is "corrected" (in Yin *et al.* [3]) for the location of the tensile ligament and the use of rheological model for deformation by Hart [6, 7].

Landes and Wei [2] showed that subcritical crack growth can occur in a high-strength steel under statically applied loads, even in an "inert" environment (namely, 99.9995 percent purity argon) at room temperature. Figure 6.2 demonstrates that crack undergoes a transient period of growth, with a growth rate that decayes to a slower steady-state growth. The growth rate then increases with crack prolongation under constant load, but increasing K_I that is commensurate with crack growth at constant load. Crack growth rate versus K_I data at five different temperatures, from 24 to 140°C (or 297 to 413 K) are shown in Fig. 6.3. Crack growth tests at these temperatures showed that the growth rates depended strongly on temperature, and showed a K_I-dependent apparent activation energy of about 46 to 75 kJ/mol., Fig. 6.4. The activation energy for steady-state creep deformation over this range of temperatures generally fell in the range of 50 to 84 kJ/mol.

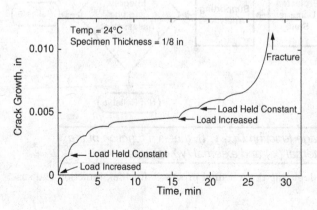

Figure 6.2. Schematic diagram of the environmental control system [2].

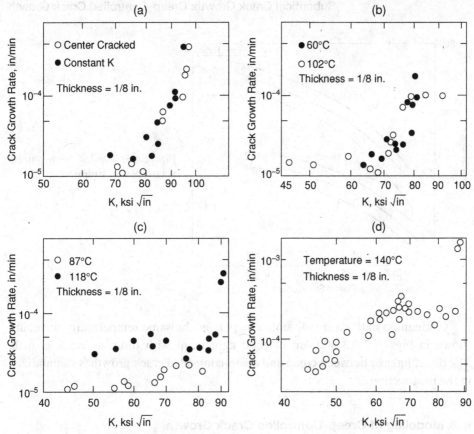

Figure 6.3. Steady-state crack growth kinetics for AISI 4340 steel in dehumidified argon [2].

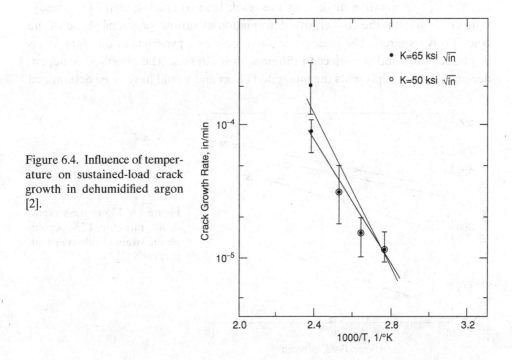

Figure 6.4. Influence of temperature on sustained-load crack growth in dehumidified argon [2].

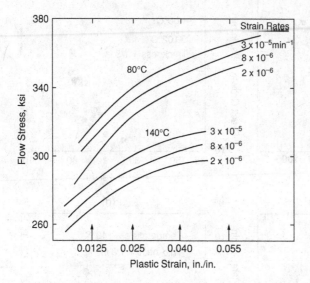

Figure 6.5. Typical flow stress versus plastic strain curve [2].

Companion data on steady-state creep, over the same temperature range, are shown in Figs. 6.5–6.8 and are in good agreement with those of crack growth. The direct linkage between creep and creep-controlled crack growth is summarized in the next sections.

6.3 Modeling of Creep-Controlled Crack Growth

The foregoing experimental observations strongly suggested the connection between creep deformation at or near the crack tip and crack growth. For steady-state crack growth, the cooperative deformation at various positions ahead of the crack tip is required. The material at these positions experiences different levels of plastic strain and is subject to different flow stresses. The observed K dependence, therefore, represents the integrated effect and would have to be determined

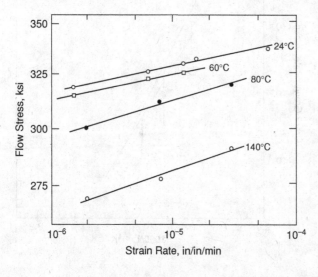

Figure 6.6. Flow stress versus strain rate for 1.25 percent plastic strain at different temperatures [2].

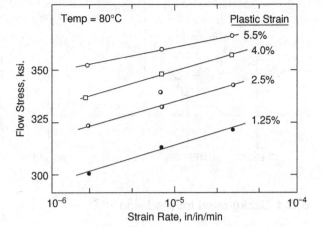

Figure 6.7. Flow stress versus strain rate for different plastic strains at constant temperature [2].

by considering the dependence of creep rate on both flow stress and structure (plastic strain) in relation to K. Even then, the K dependence for crack growth can only be related to the dependence of creep rate on flow stress through a suitable model. Here, the model by Yin *et al.* [3] is summarized. This model is a modification of the one presented by Landes and Wei [2], and explicitly incorporates a model for creep deformation.

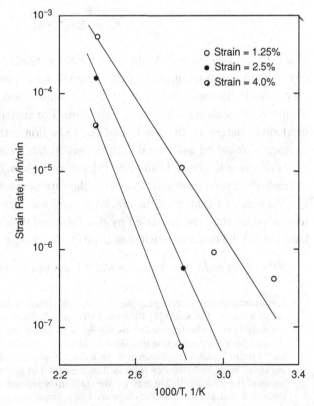

Figure 6.8. Influence of temperature on strain rates at a constant flow stress of 312 ksi [2].

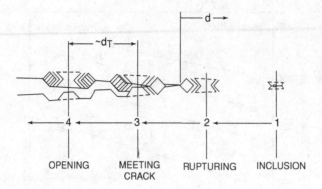

Figure 6.9. A model for the interception of an inclusion-started void by the crack front resulting in dimple formation (after Krafft [4, 5]).

6.3.1 Background for Modeling

During the late 1960s, Krafft [4] proposed a model to relate the fracture behavior of metals to its uniaxial deformation characteristics. The model is based on the concept of deformation and rupture of tensile ligaments, with the dimension of a characteristic process zone d_T, at the crack tip (see Fig. 6.9). The initial effort was described in Chapter 5. This model and its subsequent extensions [2–5] have been grouped under the title of "tensile ligament instability" (TLI) models. As discussed in Chapter 5, it was first applied to establish a relationship between the plane strain fracture toughness K_{Ic} of a material and its strain-hardening exponent, which led to the following relationship (see Chapter 5):

$$K_{Ic} = En\sqrt{2\pi d_T} \tag{6.1}$$

Krafft [4] and Krafft and Mulherin [5] later extended the TLI model to describe stress corrosion crack growth. Crack growth was viewed in terms of the instability of tensile ligaments where their lateral contraction was augmented by uniform chemical dissolution of the tensile ligaments. For sustained-load crack growth in an inert environment, on the other hand, the reduction in the cross-sectional area of the ligaments would be associated with the creep rate (Landes and Wei [2], Yin et al. [3]). Following Krafft and Mulherin [5], one considers, instead, the rate of evolution of crack-tip creep strain with respect to the rate of crack growth.[1]

Consider a ligament at the crack tip, with a current cross-sectional area A and true stress σ, the load P carried by this ligament would be $P = \sigma A$. At maximum load (i.e., at the onset of tensile instability), the change in load would be zero; i.e.,

$dP = \sigma dA + d\sigma A = 0$ at maximum load, or point of tensile instability

---

[1] Unfortunately, a subtle change in the location of the crack-tip coordinates was made in Yin et al. [3] vis-à-vis that used by Krafft [4], Krafft and Mulherin [5], and Landes and Wei [2] that affected internal consistency. In both formulations, the ligament dimension d_T is associated with the "average" distance between neighboring inclusions, or inclusion-nucleated voids. In the original formulation, the "current" crack tip is assumed to be located approximately half-way between the inclusion-nucleated voids, and attention is focused on the center of the uncracked ligament at a distance d_T ahead of the crack tip. In Yin et al. [3], the crack tip is assumed to be at the center of the inclusion-nucleated void, and the center of the uncracked ligament is now located at a distance $d_T/2$ ahead of the crack tip. Here, the Yin et al. [3] formulations are corrected to restore consistency.

For time-independent, power-hardening material (no creep):

$$\frac{d\sigma}{\sigma} = -\frac{dA}{A} = 2\upsilon d\varepsilon = d\varepsilon \text{ for } \upsilon = 1/2, \quad \text{or} \quad \frac{d\sigma}{d\varepsilon} = \sigma$$

$$\sigma = k\varepsilon^n$$

$$\frac{d\sigma}{d\varepsilon} = nk\varepsilon^{(n-1)} = \frac{nk\varepsilon^n}{\varepsilon} = \frac{n\sigma}{\varepsilon} \quad \text{or} \quad \frac{n}{\varepsilon} = 1$$

Similarly, for time-dependent, power-hardening material (*i.e.*, one that creeps), on the other hand,

$$\frac{d\sigma}{\sigma} = -\frac{dA}{A} = 2\upsilon d\varepsilon + 2\upsilon \dot{\varepsilon}_s dt = d\varepsilon + \dot{\varepsilon}_s dt \quad \text{or} \quad \frac{1}{\sigma}\frac{d\sigma}{d\varepsilon} = 1 + \frac{\dot{\varepsilon}_s}{\dot{\varepsilon}}$$

$$\sigma = k\varepsilon^n$$

$$\frac{1}{\sigma}\frac{d\sigma}{d\varepsilon} = \frac{nk\varepsilon^{(n-1)}}{k\varepsilon^n} = \frac{n}{\varepsilon} \quad \text{or} \quad \frac{n}{\varepsilon} = 1 + \frac{\dot{\varepsilon}_s}{\dot{\varepsilon}}$$

Here, creep of the ligaments enhances the evolution of strain, and leads to earlier onset of tensile ligament instability; hence, crack growth.

Extending on the concepts of Krafft [4, 5], and Landes and Wei [2], an analytical model was proposed by Yin *et al.* [3] to explore the crack growth response over a broader range of K levels. In this model, the phenomenological model of creep proposed by Hart [6] is used.

6.3.2 Model for Creep

Hart and coworkers developed a phenomenological theory of plastic deformation by using the concept of equation of state [6, 7]. The proposed deformation model consists essentially of two parallel branches (Fig. 6.10). Branch I represents

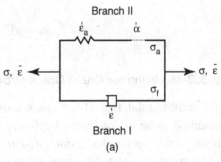

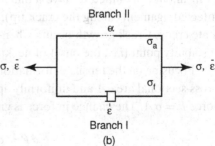

Figure 6.10. Schematic rheological diagram representing Hart's deformation model [6, 7].

dislocation-glide-controlled processes, while branch II depicts diffusion-controlled processes. In the model the α element represents the barrier processes, the $\dot{\varepsilon}_a$ element characterizes the pile-ups as stored strain, and the $\dot{\varepsilon}$ element represents glide friction. The applied uniaxial tensile stress σ is the sum of the stresses σ_a and σ_f that operate in each branch. Since $\dot{\varepsilon}_a$ is generally small in comparison with $\dot{\alpha}$ for high-strength alloy, it is reasonable to simplify the model by neglecting $\dot{\varepsilon}_a$ (see Fig. 6.10(b)).

Deformation is obviously controlled by both the dislocation glide processes and the diffusive processes. The contribution from each process may be more, or less, at different temperatures. Hence, both branches of the phenomenological model will operate such that:

$$\sigma = \sigma^* \exp\left\{-\left(\frac{\dot{\varepsilon}^*}{\dot{\varepsilon}_s}\right)^{\lambda}\right\} + G\left(\frac{\dot{\varepsilon}_s}{\dot{A}^*}\right)^{1/M} \tag{6.2}$$

where λ and M are material constants with typical values of $\lambda = 0.15$ and $M = 7$–9, G is the shear modulus, σ^* is the hardness, $\dot{\varepsilon}^*$ and $\dot{A}^*$ are strain rate coefficients, and $\dot{\varepsilon}_s$ $(-\dot{\varepsilon})$ is the steady-state creep rate. This equation cannot be simplified and is too complex to yield an expression for the steady-state creep rate $\dot{\varepsilon}_s$.

At temperatures below the homologous temperature ($T_{\text{homo}} \approx 0.25 T_{\text{melt}}$), however, the deformation processes are predominantly controlled by dislocation glide; i.e., $\dot{\varepsilon}^* \to 0$ [6, 7]. The lower branch in Fig. 6.10(b), therefore, is more important and the deformation rate is well described by transforming Eqn. (6.2):

$$\sigma = \sigma^* \exp\left\{-\left(\frac{\dot{\varepsilon}^*}{\dot{\varepsilon}_s}\right)^{\lambda}\right\} + G\left(\frac{\dot{\varepsilon}_s}{\dot{A}^*}\right)^{1/M} \quad \Rightarrow \quad \sigma = \sigma^* + G\left(\frac{\dot{\varepsilon}_s}{\dot{A}^*}\right)^{1/M}$$

Hence,

$$\dot{\varepsilon}_s = \dot{A}^*\left(\frac{\sigma - \sigma^*}{G}\right)^{M} \tag{6.3}$$

6.3.3 Modeling for Creep Crack Growth

To simplify modeling of the crack growth rate, the material near the crack tip is assumed to be in the form of cylindrical tensile microligaments that have formed from "voids" generated from surrounding inclusion particles. For simplicity, each of ligament is assumed to have a diameter d_T (representing the average size of the affected ligaments along the crack tip), is homogeneous, and is subjected to steady-state creep. Crack growth occurs when the ligaments reach the tensile deformation instability point (i.e., the onset of necking).

Focusing on the tensile deformation instability of a (noncreeping) ligament with cross-sectional area A and uniformly applied stress (on average) σ, and the applied force $P = \sigma A$. The change in force is given by:

$$dP = \sigma dA + d\sigma A \tag{6.4}$$

At the maximum load, the change in load is zero (*i.e.*, $dP = 0$) such that, for a time-independent, power-hardening material (*i.e.*, one that does not creep):

$$\frac{d\sigma}{\sigma} = -\frac{dA}{A} = 2vd\varepsilon = d\varepsilon \quad \text{or} \quad \frac{d\sigma}{d\varepsilon} = \sigma; \quad (v = 0.5 \text{ for constant volume deformation})$$

$$\because \sigma = k\varepsilon^n \tag{6.5}$$

$$\frac{d\sigma}{d\varepsilon} = nk\varepsilon^{(n-1)} = \frac{nk\varepsilon^n}{\varepsilon}, \quad \text{or} \quad \frac{n}{\varepsilon} = 1$$

In other words, the strain at the onset of tensile deformation instability (maximum load point) is equal to the strain-hardening exponent. For a time-dependent, power-hardening material (*i.e.*, one that creeps), on the other hand, deformation is enhanced by creep, such that:

$$\frac{d\sigma}{\sigma} = -\frac{dA}{A} = 2vd\varepsilon + 2v\dot{\varepsilon}_s dt = d\varepsilon + \dot{\varepsilon}_s dt \quad \text{or} \quad \frac{1}{\sigma}\frac{d\sigma}{d\varepsilon} = 1 + \frac{\dot{\varepsilon}_s}{\dot{\varepsilon}}; \quad (v = 0.5)$$

$$\because \quad \sigma = k\varepsilon^n \tag{6.6}$$

$$\frac{d\sigma}{d\varepsilon} = nk\varepsilon^{(n-1)} = \frac{nk\varepsilon^n}{\varepsilon}, \quad \text{or} \quad \frac{n}{\varepsilon} = 1 + \frac{\dot{\varepsilon}_s}{\dot{\varepsilon}}$$

whereby, the tensile deformation strain is enhanced by steady-state creep.

According to Landes and Wei [2], the connection between the steady-state creep rate and the crack-driving force (characterized by K) is derived through the use stress-strain results from elastic-plastic analysis by Hutchinson [9] and Rice and Rosengren [10]. According to these models, crack-tip stress and strains in the loading direction (*y*-direction) are given by Eqn. (6.7).

$$\sigma_{yy}(r, 0) = 1.2\sigma_{ys}\left(\frac{K}{\sigma_{ys}\pi^{1/2}}\right)^{\frac{2}{N+1}} r^{-\frac{1}{N+1}} \tag{6.7 a}$$

$$\varepsilon_{yy}(r, 0) = 0.75\varepsilon_{ys}\left(\frac{K}{\sigma_{ys}\pi^{1/2}}\right)^{\frac{2N}{N+1}} r^{-\frac{N}{N+1}} \tag{6.7 b}$$

Differentiating strain (Eqn. (6.7b)) with respect to time, the strain rate at the position $(r, 0)$ ahead of the crack tip becomes:

$$\dot{\varepsilon}_{yy}(r, 0) = 0.75\varepsilon_{ys}\frac{2N}{N+1}\left(\frac{K}{\sigma_{ys}\pi^{1/2}}\right)^{\frac{2N}{N+1}}\frac{\dot{K}}{K}r^{-\frac{N}{N+1}}$$

$$- 0.75\varepsilon_{ys}\frac{N}{N+1}\left(\frac{K}{\sigma_{ys}\pi^{1/2}}\right)^{\frac{2N}{N+1}} r^{-\frac{N}{N+1}}\frac{\dot{r}}{r} \tag{6.8}$$

Evaluating the derivative at $r = d_T$ (*i.e.*, at the center of the unbroken ligament immediately ahead of the crack tip[2]), Eqn. (6.8) becomes:

$$\dot{\varepsilon}_{yy}(d_T, 0) = 0.75\varepsilon_{ys}\frac{2N}{N+1}\left(\frac{K}{\sigma_{ys}\pi^{1/2}}\right)^{\frac{2N}{N+1}} d_T^{-\frac{N}{N+1}}\frac{\dot{K}}{K}$$

$$- 0.75\varepsilon_{ys}\frac{N}{N+1}\left(\frac{K}{\sigma_{ys}\pi^{1/2}}\right)^{\frac{2N}{N+1}} d_T^{-\frac{N}{N+1}}\frac{\dot{r}}{d_T} \tag{6.9}$$

---

[2] The choice of $r = d_T$ brings the choice of coordinates in the Yin *et al.* [3] model into conformance with that of Krafft [4], Krafft and Mulherin [5], and Landes and Wei [2].

It may be seen that the term $N/(N+1)$ is nearly equal to 1. As such, the relative contributions to the strain rate are determined by the ratios $\dot{K}/K$ and $\dot{r}/d_T$. It may be readily shown that:

$$\frac{\dot{K}}{K} \to \frac{\dot{a}}{a} \quad \text{and} \quad -\frac{\dot{r}}{d_T} = \frac{\dot{a}}{d_T}.$$

Because the crack length a is much larger than the ligament (or process zone) size d_T, the second of the two terms in Eqn. (6.9) dominates. By setting $\dot{a} = -\dot{r}$, Eqn. (6.9) becomes

$$\dot{\varepsilon}_{yy}(d_T, 0) \approx 0.75\varepsilon_{ys} \frac{N}{N+1} \left(\frac{K}{\sigma_{ys}\pi^{1/2}} \right)^{\frac{2N}{N+1}} d_T^{-\frac{2N}{N+1}} \frac{\dot{a}}{d_T}, \tag{6.10}$$

The process zone size d_T is estimated from Eqn. (6.7b) by setting $r = d_T$, $\varepsilon = 1/N$, and $K = K_c$ and solving for $r (= d_T)$:

$$d_T = \left(0.75N\varepsilon_{ys}\right)^{\frac{N+1}{N}} \left(\frac{K_c}{\sigma_{ys}\pi^{1/2}} \right)^2 \tag{6.11}$$

By substituting Eqn. (6.11) into Eqn. (6.7b), the values of strain at $r = d_T$ for any value of K may be obtained.

$$\varepsilon_{r(d_T,0)} = \frac{1}{N} \left(\frac{K}{K_c} \right)^{\frac{2N}{N+1}} \tag{6.12}$$

By combining Eqns. (6.6) and (6.12), the rate of steady-state creep crack growth is related to the steady-state creep rate and other measurable properties of the material; namely [3],

$$\dot{a} = \frac{da}{dt} = \frac{(N+1)d_T}{\left[1 - \left(\frac{K}{K_c} \right)^{\frac{2N}{N+1}} \right]} \dot{\varepsilon}_s (r = d_T) \tag{6.13}$$

For temperatures well below the homologous temperature, the steady-state creep rate is well represented by Eqn. (6.3). By substituting Eqn. (6.3) into Eqn. (6.13), the steady-state creep crack growth rate becomes:

$$\dot{a} = \frac{da}{dt} = \frac{(N+1)d_T}{\left[1 - \left(\frac{K}{K_c} \right)^{\frac{2N}{N+1}} \right]} \dot{A}^* \left(\frac{\sigma - \sigma^*}{G} \right)^M \tag{6.14}$$

The quantities N, M, A^*, σ^*, and G are determined independently from uniaxial deformation tests, and K_c is determined from fracture toughness tests. The process zone size d_T may be estimated metallographically from polished or fractured specimens. In other words, no empirical fitting parameters are involved in this relationship between crack growth and creep deformation rates.

The effect of local strain near the crack tip on crack growth is reflected through the hardness parameter σ^*, which is only a function of strain level [3]. For simplicity in calculating crack growth rates, σ^* is assumed to be constant and its value at a strain level of $\varepsilon = n$ is used.

6.4 Comparison with Experiments and Discussion

Equation 6.14 provides a formal connection between creep crack growth and the kinetics of creep deformation in that the steady-state crack growth rates can be predicted from the data on uniaxial creep deformation. Such a comparison was made by Yin et al. [3] and is reconstructed here to correct for the previously described discrepancies in the location of the crack-tip coordinates (from $d_T/2$ to d_T) with respect to the microstructural features, and in the fracture and crack growth models. Steady-state creep deformation and crack growth rate data on an AISI 4340 steel (tempered at 477 K), obtained by Landes and Wei [2] at 297, 353, and 413 K, were used. (All of these temperatures were below the homologous temperature of about 450 K.) The sensitivity of the model to σ_{ys}, N, and σ^* is assessed.

6.4.1 Comparison with Experimental Data

A formal linkage of crack growth and creep deformation kinetics were made by Yin et al. [3], and is summarized here. Here, the Yin et al. model is adjusted to conform the "tensile ligament" location, with respect to the crack tip, to that of the original Krafft [4] and Krafft and Malherin [5] model for fracture. The original uniaxial creep deformation and steady-state crack growth data from a single lot of AISI 4340 steel, at 297, 353, and 413 K (all below the homologous temperature of about 450 K) are summarized in [2]. Comparisons of the model against the test data are summarized here. The impact of the coordinate location adjustment is discussed.

The required input material parameters are as follows (see Yin et al. [3]): σ^*, M, $\dot{A}^*$, σ_{ys}, N, K_c, and d_T. The hardness σ^* values at different strain levels and the strain rate exponent M were obtained empirically and are shown in Table 6.2 [3]. The values of σ^* were obtained from a best-fit curve of σ^* vs. ε results at $\varepsilon = n$ (see, for example, Fig. 6.11). Values for other pertinent variables were derived or estimated by Yin et al. [3] from other sources, and are summarized in Table 6.2.

The predicted steady-state crack growth kinetics (conforming to the originally defined crack-tip location by Krafft [4] and Krafft and Mulherin [5]), based on the parameters given in Table 6.3, are shown in Figs. 6.12 to 6.14 as solid curves in comparison with the experimental data from Landes and Wei [2], at 297, 353, and 413 K respectively. Agreement over the range of available data is improved, and reflects the factor of 2 "correction" in ligament location (namely, d_T) and the concomitant changes in strain level and flow stress. Conformance with the data trend affirms the concept of creep-controlled crack growth. It suggests that the model might apply over a broader range of K levels, but needs to be confirmed. Also, other models

Table 6.2. *Hardness σ^* and material constant M* [3]

$\varepsilon(\%)$	1.25	2.50	4.00	5.50
σ^*	1779	1875	1960	2149
M	7.62	7.62	7.62	7.62

Table 6.3. *Parameters needed for modeling* [3]

T(K)	σ_{ys} (MPa)	K_c (MPa-m$^{1/2}$)	σ^*(MPa)	N	d_T (μm)	$\dot{A}$ (s^{-1})	M
297	1447	108	2010	9.5	16.1	2.0×10^{10}	7.62
353	1378	104	1985	10.5	16.1	6.7×10^{11}	7.62
413	1323	103	1948	12.5	16.1	9.0×10^{14}	7.62

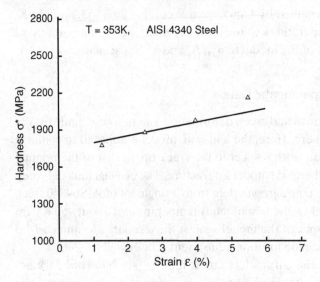

Figure 6.11. Hardness (σ^*) versus strain (ε) for AISI 4340 steel [3].

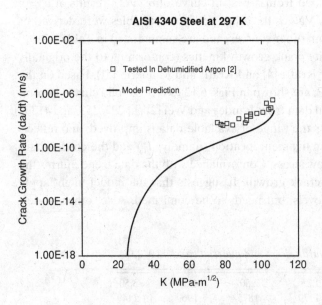

Figure 6.12. Comparison of "corrected" model prediction with crack growth data on AISI 4340 steel at 297 K [3].

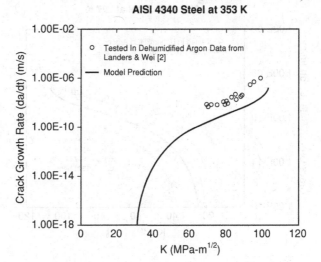

Figure 6.13. Comparison of "corrected" model prediction with crack growth data on AISI 4340 steel at 353 K [3].

for creep deformation may have to be considered for other materials and for crack growth response over other ranges of temperatures.

6.4.2 Model Sensitivity to Key Parameters

Yin *et al.* [3] examined the sensitivity of the model to certain key parameters, such as σ_{ys}, N, and σ^*. The predicted crack growth rates were found to be very sensitive to the yield strength (σ_{ys}) and strain-hardening exponent ($n = 1/N$); see Figs. 6.15 and 6.16, where the curves (shown by the solid lines) were calculated on the basis of parameters given in Table 6.2. For a given value of N, a 5% increase in yield strength resulted in a tenfold increase in the predicted crack growth rates (see Fig. 6.15). At

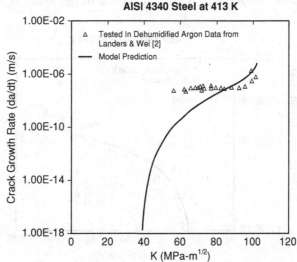

Figure 6.14. Comparison of "corrected" model prediction with crack growth data on AISI steel at 413 K [3].

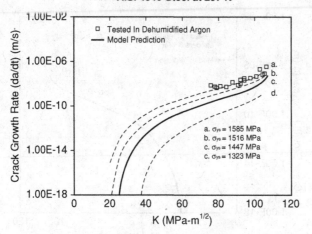

Figure 6.15. Sensitivity of model prediction to yield strength σ_{ys} [3].

constant σ_{ys}, on the other hand, a 10% increase in N (*i.e.*, 10% softening) decreases the predicted crack growth rates by an order of magnitude (or tenfold) (Fig. 6.16). The actual dependence of crack growth rates on these parameters, as well as yield strength, remains to be verified by further experiments.

Of the three parameters, σ^* is the most difficult to estimate and is perhaps the least certain to estimate. The estimated influence of hardness σ^* on crack growth rate is shown in Fig. 6.17, and reflects the effect of local strain ε ahead of the crack tip. Examination of Fig. 6.17 suggests that a 10% reduction in σ^* from 2010 to about 1800 MPa could conform the creep crack growth rate model to the experimental data.

Because crack-tip strain increases with K, σ^* is expected to increase (see Fig. 6.11) commensurately and to alter crack growth response. Its potential influences have been estimated by Yin *et al.* [3], and are shown as dashed curves in Figs. 6.13, 6.14 and 6.17 (using the revised data). Even with this "weak" dependence of σ^* on K, the resulting influence in the lower K region is quite large. Further

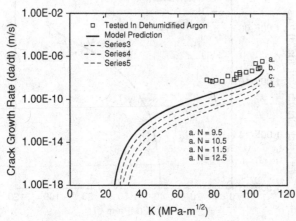

Figure 6.16. Sensitivity of model prediction to inverse strain-hardening exponent N [3].

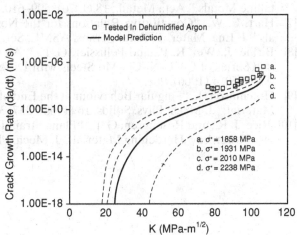

Figure 6.17. Sensitivity of model prediction to hardness σ^* [3].

experiments, over a broader range of materials and conditions, are needed to better characterize and understand this phenomenon. As suggested by Yin *et al.* [3], more suitable, and a broader range of, materials should be considered.

6.5 Summary Comments

The occurrence of creep-controlled crack growth, in an inert environment, has been demonstrated. It can occur even at modest temperatures, and has been linked to localized creep deformation and rupture of ligaments isolated by the growth of "inclusion-nucleated" voids ahead of the crack tip. Landes and Wei [2] and Yin *et al.* [3] have made a formal connection between the two processes, and provided a modeling framework and experimental data to link the kinetics of creep to creep-controlled crack growth. Further work is needed to develop, validate, and extend this understanding. In particular, its extension to high-temperature applications needs to be explored.

REFERENCES

[1] Li, C. Y., Talda, P. M. and Wei, R. P., unpublished research. Applied Research Laboratory, U.S. Steel Corp., Monroeville, PA (1966).
[2] Landes, J. D., and Wei, R. P., "Kinetics of Subcritical Crack Growth and Deformation in a High Strength Steel," J. Eng'g. Materials and Technology, Trans. ASME, Ser. H, 95 (1973), 2–9.
[3] Yin, H., Gao, M., and Wei, R. P., "Deformation and Subcritical Crack Growth under Static Loading," J. Matls. Sci. & Engr., A119 (1989), 51–58.
[4] Krafft, J. M., "Crack Toughness and Strain Hardening of Steels," Applied Materials Research, 3 (1964), 88–101.
[5] Krafft, J. M., and Mulherin, J. H., "Mechanical Behavior of Materials," Vol. 2, Proc. ICM3, Cambridge, U.K., Pergamon, Oxford (1979), 383–396.

[6] Hart, E. W., "A Phenomenological Theory for Plastic Deformation of Polycrys-
     talline Metals," Acta Metall., 18 (1970), 599–610.

[7] Hart, E. W., "Constitutive Relations for the Nonelastic Deformation of Met-
     als," J. Eng. Mater. Technol., Trans. ASME, Ser. H., 98 (1976), 193–202.

[8] Birkle, J., Wei, R. P., and Pellissier, G. E., "Analysis of Plane-Strain Fracture
     in a Series of 0.45C-Ni-Cr−Mo Steels with Different Sulfur Contents," Trans.
     ASM, 59, 4 (1966), 981.

[9] Hutchinson, J., "Singular Behaviour at the End of a Tensile Crack in Hardening
     Material," J. Mech. Phys. Solids, 16 (1968), 337–342.

[10] Rice, J. R., and Rosengren, G. F., "Plane Strain Deformation Near a Crack Tip
     in a Power Law Hardening Material," J. Mech. Phys. Solids, 16 (1968), 1–12.

7 Subcritical Crack Growth: Stress Corrosion Cracking and Fatigue Crack Growth (Phenomenology)

7.1 Overview

Stress corrosion, or stress corrosion cracking (SCC), and fatigue/corrosion fatigue, or fatigue crack growth (FCG), are problems of long standing. They manifest themselves in the occurrence of "delayed failure" (*i.e.*, failure after some period of time or numbers of loading cycles) of structural components under statically or cyclically applied loads, at stresses well below the yield strength of the material. These phenomena of delayed failure are often referred to as "static fatigue," for SCC, or simply "fatigue" for cyclically varying loads. The traditional measure of stress corrosion cracking susceptibility is given in terms of the time required to produce failure (time-to-failure) at different stress levels, as obtained from testing "smooth" or "notched" specimens of the material in the corrosive environments (for example, sea water, for marine applications). For fatigue, the measure is given by the number of cycles to cause failure (the fatigue life) at given cyclic stress levels, or the endurance limit (stress corresponding to some prescribed number of load cycles; *e.g.*, 10^6 cycles).

The failure time, however, incorporates both the time required for "crack initiation" and a period of slow crack growth so that the separate effect of the environment on each of these stages cannot be ascertained. (Some of the difficulty stems from the lack of a precise definition for crack initiation.) This difficulty is underscored by the results of Brown and Beachem [1] on SCC of titanium alloys. They showed that certain of the alloys that appeared to be immune to stress corrosion cracking in the traditional (smooth specimen) tests are, in fact, highly susceptible to environment-enhanced crack growth. The apparent immunity was explained by the fact that these alloys were nearly immune to pitting corrosion, which was required for crack nucleation in the same environment [1].

Prior to the 1960s, stress corrosion cracking and corrosion fatigue were principally under the purview of corrosion chemists and metallurgists, and the primary emphasis was on the response of materials in aqueous environments (*e.g.*, sea/salt water), particularly for SCC because of the relative ease of experimentation. Much of the attention was devoted to the understanding of electrochemical reactions that are associated with metal dissolution, crack nucleation, and time-to-failure under a constantly applied load or strain (using smooth or notched specimens) in the corrosive environment. In simplest terms, the chemical/electrochemical processes involve (i) metal oxidation/dissolution, (ii) the dissociation of water, (iii) the formation of metal hydroxide, and (iv) hydrogen reduction. Specifically, the elementary reactions involved in the dissolution of a metal with a valance of n are represented by the following half-reactions (here, for water) (namely, the *anodic* metal oxidation and the *cathodic* hydrogen reduction reactions):

$$M \rightarrow M^{+n} + ne^-$$
$$nH_2O \rightarrow nH^+ + nOH^-$$
$$M^{+n} + nOH^- \rightarrow M(OH)_n$$
$$nH^+ + ne^- \rightarrow \frac{n}{2}H_2 \uparrow$$

The corrosion/stress corrosion communities, believing that cracking is the result of localized metal dissolution at the crack tip, focused on the evolution of crack-tip chemistry and the anodic part of these coupled electrochemical reactions in the understanding and control of SCC and CF. Others, including this author, on the other hand, believe that the hydrogen that evolves through these reactions can enter the material at the crack tip, and is directly responsible for enhanced cracking; *albeit*, the overall rate of crack growth may be controlled by coupled electrochemical reactions. From the design/engineering perspective, however, emphasis was placed on the establishment of allowable design threshold stresses that would provide assurance of "safety" over the design life of the component/structure.

With its development and usage since the late 1950s, driven principally by the aerospace and naval programs at the time, fracture mechanics has become the principal framework for engineering design and for fundamental understanding of materials response. For SCC, emphasis shifted to the use of fracture mechanics parameters to characterize stress corrosion-cracking thresholds (namely, K_{Iscc}) and crack growth kinetics in terms of the dependence of crack growth rate (da/dt) on the driving force, now characterized by K_I. For fatigue, the author will principally draw on his own and his coworkers' experience and research to provide an overview of a segment of this field in the following chapters on subcritical crack growth; namely, stress corrosion and fatigue crack growth. The reader is encouraged to examine the extensive literature by others to obtain a broader perspective of the field.

7.2 Methodology

Before delving into the topic, it is important to prescribe the intent of this chapter. Here, the methodology used in assessing stress corrosion/sustained-load crack growth and fatigue/corrosion fatigue crack growth is highlighted. The methodology is intended for the measurement of (or is presumed to be measuring) steady-state response and its use in structural life estimation and management. It would reflect the conjoint actions of mechanical loading and chemical/electrochemical

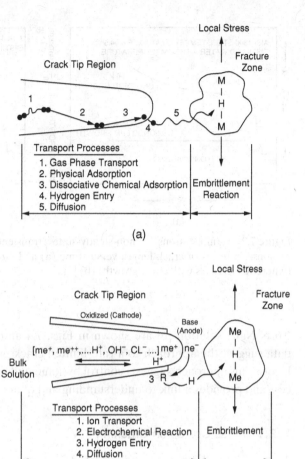

Figure 7.1. Schematic illustration of the sequential processes for environmental enhancement of crack growth by gaseous (a) and aqueous (b) environments. Embrittlement by hydrogen is assumed, and is schematically depicted by the metal-hydrogen-metal bond.

interactions with the material's microstructure (see Fig. 7.1). Only terms of steady-state crack growth are considered herein. Non-steady-state behaviors (that reflect, for example, the evolution of "steady-state" crack-front shape and crack-tip chemical/electrochemical environment) are counted as incubation, at times in excess of several thousand hours [2, 3]. The presence of these non-steady-state responses is illustrated in Figs. 7.2 and 7.3. Similar transient responses have been observed for fatigue crack growth and are treated similarly [4].

Figure 7.2a shows the evolution of crack growth in distilled water at two starting K_I or load levels, both showing an initially rapid stage of transient crack growth that quickly decayed, coming to an apparent arrest at the lower load, and accelerating growth in the other. The presence of these non-steady-state responses are reflected in the "steady-state" da/dt vs. K_I plots by the "tails," or "false thresholds," in Fig. 7.2b. The influence of temperature and K level on these non-steady-state responses is illustrated in Figs. 7.3a and 7.3b. Representative sustained-load crack growth data on a Ti-5Al-2.5Sn in hydrogen [5], and on an AISI 4340 steel in

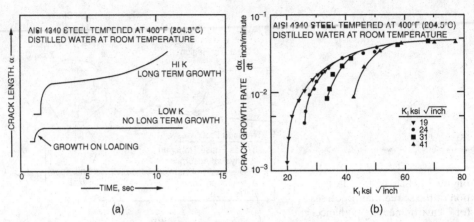

Figure 7.2. Manifestations of non-steady-state (transient) and steady-state crack growth response in terms of crack length versus time (a) and *da/dt* versus K_I under constant load (where *K* increases with crack growth) (b) [3].

0.6 N NaCl solution [6] are shown in Figs. 7.4 and 7.5, respectively. Both sets of data suggest the approach to a rate-limited crack growth over a broad range of *K* levels. This approach suggests control by some underlying reaction or transport process and provides a link to understanding and quantification of response.

7.2.1 Stress Corrosion Cracking

The overall SCC response is illustrated diagrammatically in Fig. 7.6, with the rate-limited stage of crack growth represented by stage II (left-hand figure) and a schematic representation of the influence of "incubation" (on the right). From a

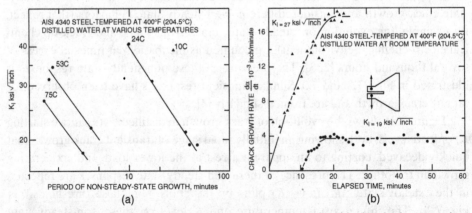

Figure 7.3. Typical sustained-load crack growth response, showing incubation, transient (non-steady-state) and steady-state crack growth, under constant load (where *K* remained constant, with crack growth, through specimen contouring) [3].

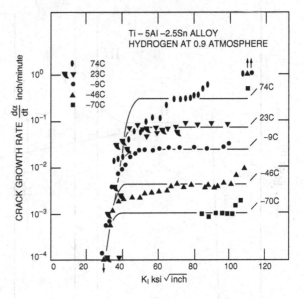

Figure 7.4. Typical kinetics of sustained-load crack growth for a Ti-5Al-2.5Sn in gaseous hydrogen at 0.9 atm and temperatures from 223 to 344 K (−70 to 74C) [5].

design perspective, the contribution to life is estimated from the crack growth portion as follows:

$$\frac{da}{dt} = F(K_I, \text{environ.}, T, \text{etc.})$$

$$t_{SC} = \int_{a_i}^{a_f} [F(K_I, \text{environ.}, T, \text{etc.})]^{-1} da \tag{7.1}$$

The functional relationship between K_I and a depends on geometry and loading, and is assumed to be known. As such:

$$\therefore \frac{dK_I}{dt} = \frac{dK_I}{da}\frac{da}{dt} = \frac{dK_I}{da}F(K_I, \text{environ.}, T, \text{etc.})$$

$$\therefore t_{SC} = \int_{K_{Ii}}^{K_{Ic}} \left[\frac{dK_I}{da}F(K_I, \text{environ.}, T, \text{etc.})\right]^{-1} dK_I \tag{7.2}$$

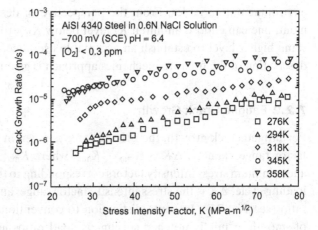

Figure 7.5. Typical kinetics of sustained-load crack growth for an AISI 4340 steel in 0.6 N NaCl solution at temperatures from 276 to 358 K [6].

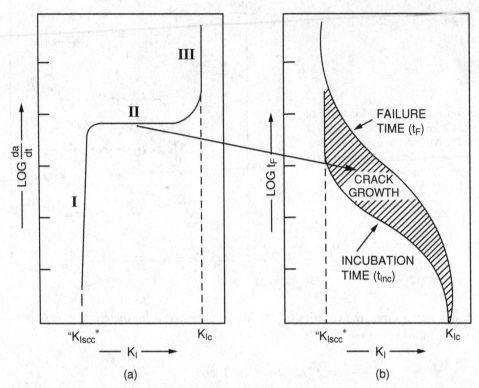

Figure 7.6. Typical sustained-load (stress corrosion) cracking response in terms of steady-state crack growth rates (left) and time (right) [3].

where the change in K with crack growth (dK/da) depends on geometry. The time-to-failure, t_F, is then:

$$t_f = t_{INC}(K_{I_i}, \text{environ.}, T, \text{etc.}) + t_{SC}(K_I, \text{geometry, environ., } T, \text{etc.}) \quad (7.3)$$

The crack growth contribution to t_F is depicted by the shaded area in Fig. 7.6, and the contribution by "incubation" is schematically indicated by the clear region in Fig. 7.6b. The missing key information is the functional dependence of the SCC crack growth rate on the the crack-driving force K_I, and the relevant material and environmental variables. From the perspective of design, or service life management, one can choose an initial K_I level below K_{Iscc} to "achieve" indefinite life, or some higher level to establish an acceptable useful/economic life. A more detailed description of the fracture mechanics approach is given in Wei [2].

7.2.2 Fatigue Crack Growth

For fatigue crack growth, the driving force is given in terms of the stress intensity factor range; namely, $\Delta K = K_{max} - K_{min}$, where K_{max} and K_{min} are the maximum and minimum stress intensity factors corresponding to the respective loads in a given loading cycle, ΔK is the stress intensity factor range, and $K_{min}/K_{max} = R$ is the load ratio (see Fig. 7.7). In contradistinction to conventional fatigue, involving the use of smooth or mildly notched specimens, load ratios less than zero ($R < 0$) is not

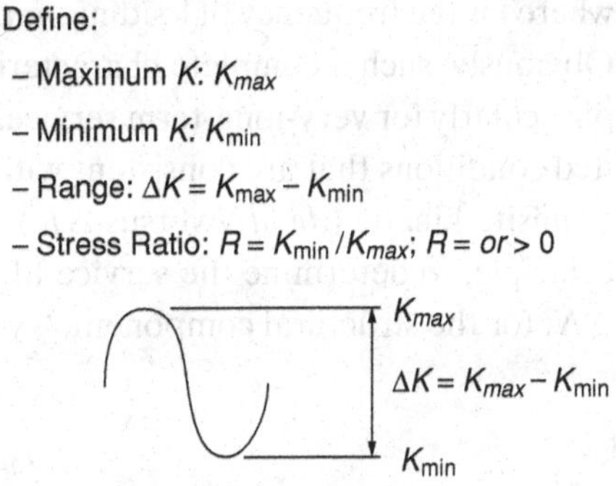

Figure 7.7. Definition of driving force parameters for fatigue crack growth.

Define:
- Maximum K: K_{max}
- Minimum K: K_{min}
- Range: $\Delta K = K_{max} - K_{min}$
- Stress Ratio: $R = K_{min}/K_{max}$; $R = or > 0$

$$\Delta K = K_{max} - K_{min}$$

considered, because compressive loading would bring the crack faces into physical contact and bring the effective driving force K_{min} to zero.

Typical crack growth rate (da/dN) versus ΔK or K_{max} curves are shown in Fig. 7.8 [4] as a function of ΔK, or K_{max}, and other loading, environmental, and material variables. Ideally, it is desirable to characterize the fatigue crack growth behavior in terms of all of the pertinent loading, material, and environmental variables, namely,

$$\frac{da}{dN} \approx \frac{\Delta a}{\Delta N} = F(K_{max}\ or\ \Delta K,\ R,\ f,\ T,\ p_i,\ C_i,\ \ldots) \tag{7.4}$$

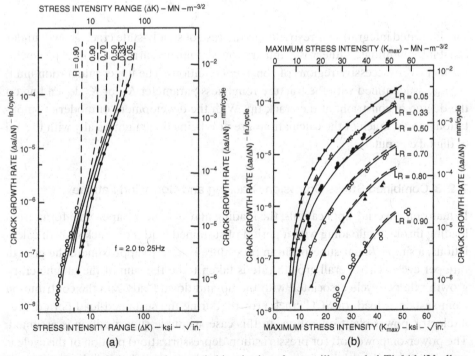

Figure 7.8. Typical fatigue crack growth kinetic data for a mill-annealed Ti-6Al-4V alloy: (a) as a function of ΔK, and (b) as a function of K_{max} [4].

where f is the frequency of loading, and $T, p_i, C_i, \ldots$ are environmental variables, etc. Obviously, such a complete characterization is not feasible and cannot be justified, particularly for very-long-term service. Data, therefore, must be obtained under limited conditions that are consistent with the intended service. Having established the requisite kinetic (da/dN versus ΔK) data, Eqn. (7.4) can be integrated, at least in principle, to determine the service life, N_F, or and appropriate inspection interval, ΔN, for the structural component.

$$N_F = \int_{a_i}^{a_f} \frac{da}{F(K_{\max}, \ldots)} \tag{7.5}$$

$$\Delta N = N_2 - N_1 = \int_{a_1}^{a_2} \frac{da}{F(K_{\max}, \ldots)} \tag{7.6}$$

The lower limit of integration (a_i or a_1) is usually defined on the basis of nondestructive inspection (NDI) capabilities, or on prior inspection; the upper limit is defined by fracture toughness or a predetermined allowable crack size that is consistent with inspection requirements (a_f or a_2). Equations (7.5) and (7.6) may be rewritten in terms of the stress intensity factor K:

$$N_F = \int_{(K_i)_{\min}}^{(K_f)_{\max}} \frac{dK}{\dfrac{dK}{da} F(K_{\max}, \ldots)} \tag{7.7}$$

$$\Delta N = N_2 - N_1 = \int_{(K_1)_{\min}}^{(K_2)_{\max}} \frac{dK}{\dfrac{dK}{da} F(K_{\max}, \ldots)} \tag{7.8}$$

The indicated integration is restricted to the case of steady-state crack growth under constant conditions. Otherwise, integration should be carried out in a piecewise manner over successive regions of constant conditions. The upper integration limit, $(K_f)_{\max}$, is identified with the fracture toughness parameter K_{Ic} or K_c depending on the degree of constraint at the crack tip. With the development of modern computational tools, much of the calculations are now being done numerically with the aid of digital computers.

7.2.3 Combined Stress Corrosion Cracking and Corrosion Fatigue

In many engineering applications, the loading may assume a trapezoidal form, from loading through unloading with a period of sustained load. For crack growth calculations, a simple linear superposition procedure is used to approximate the growth rate per cycle. The overall growth rate is taken to be the sum of the fatigue crack growth rate per cycle associated with the up-and-down loads, and the contribution from sustained load, or SCC crack growth, during the sustained-load period; *e.g.*, in electric power plant operation. In this case, a simple superposition is assumed. The power-on/power-off (or pressurization/depressurization) portion of the cycle is associated with the driving force for fatigue crack growth. The overall rate of crack growth per cycle is given by the sum of the cycle and time-dependent contributions in Eqn. (7.9) [7]:

$$\left(\frac{\Delta a}{\Delta N}\right) \approx \left(\frac{da}{dN}\right) = \left(\frac{da}{dN}\right)_{cyc} + \left(\frac{da}{dN}\right)_{tm}$$

or   (7.9)

$$\left(\frac{\Delta a}{\Delta N}\right) \approx \left(\frac{da}{dN}\right) = \left(\frac{da}{dN}\right)_{cyc} + \int_0^\tau \left[\frac{da}{dt}(K(t))\right]_{scc} dt$$

7.3 The Life Prediction Procedure and Illustrations [4]

Although life prediction appears to be straightforward in principle, the actual prediction of service life can be quite complex and depends on the ability of the designer to identify and cope with various aspects of the problem. The life prediction procedure may be broadly grouped into four parts:

1. Structural analysis: Identification of probable size and shape of cracks at various stages of growth, their location in the component, and proper stress analysis of these cracks, taking into account the crack and component geometries and the type of loading.
2. Mission profile: Proper prescription of projected service loading and environmental conditions, with due consideration of variations in actual service experience.
3. Material response: Determination of fracture toughness and characterization of fatigue crack growth response of the material in terms of the projected service loading and environmental conditions.
4. Life prediction: Synthesis of information from the previous three parts to estimate the service life of a structural component.

The process is illustrated by the following simplified examples on fatigue crack growth under constant amplitude fatigue loading. Example 1 illustrates the growth of a central, through-thickness crack in a plate, and Example 2 illustrates the growth of a semicircular surface crack or part-through crack through the plate. (Note that, for these illustrations, the functionality of the crack growth rate dependence on ΔK is assumed to be fixed; *i.e.*, the exponent n in the crack growth "law" is assumed to be constant. In reality, the value of n changes with crack growth and the concomitant increase in ΔK, as the fracture mode changes from flat to increasing amounts of shearing mode of failure (see Fig. 7.8).)

EXAMPLE 1 – THROUGH-THICKNESS CRACK. The case of a center-cracked plate, subjected to constant-amplitude loading, is considered to provide physical insight.

Structural Analysis For a through-thickness crack of length $2a$ in a "wide" plate, subjected to uniform remote tension, σ, perpendicular to the plane of the crack, the stress intensity factor K is given by $K = \sigma\sqrt{\pi a}$. It is assumed that the crack plane remains perpendicular to the tensile axis during crack growth. The initial half-crack length (a_i) is defined by nondestructive inspection (NDI). For simplicity, the stress amplitude ($\Delta\sigma$), stress ratio (R), frequency (f), temperature (T), etc., are assumed to be constant. Furthermore, R is assumed to be greater than or equal to zero (*i.e.*, no compression), and to remain constant.

Material Response – For simplicity, the kinetics of fatigue crack growth will be assumed to be describable by a single equation over the entire range of interest; *i.e.*, $da/dN = A(\Delta K)^{2n}$ for the conditions prescribed. The fracture toughness of the material is given by K_c

From the foregoing information, the following conditions are determined:

$$\sigma_{max} = \Delta\sigma/(1 - R) = \text{constant}$$

$$\sigma_{min} = R\sigma_{max} = R\Delta\sigma/(1 - R)$$

$$\Delta K = \Delta\sigma\sqrt{\pi a} \text{ (dynamic correction not needed}$$
$$\text{at conventional fatigue frequencies)}$$

$$K_{max} = \sigma_{max}\sqrt{\pi a} \tag{7.10}$$

$$a_f = \frac{K_c^2}{\pi\sigma_{max}^2}$$

and

$$da/dN = A(\Delta K)^{2n} = \pi^n A(\Delta\sigma)^{2n}a^n; (n > 1) \text{ is assumed here}$$

The fatigue life of the center-cracked plate is then obtained by straightforward integration of the foregoing rate equation:

$$N_F = \int_{a_i}^{a_f} \pi^n A (\Delta\sigma)^{2n} a^n \, da$$

$$= \frac{1}{\pi^n A(n - 1)(\Delta\sigma)^{2n}} \left[\frac{1}{a_i^{(n-1)}} - \frac{1}{a_f^{(n-1)}} \right]$$

$$= \frac{1}{\pi^n A(n - 1)(\Delta\sigma)^{2n}a_i^{(n-1)}} \left[1 - \left(\frac{a_i}{a_f}\right)^{(n-1)} \right]; \quad (n > 1) \tag{7.11}$$

Several things immediately become obvious from Eqns. (7.10) and (7.11): (i) A specific, independent failure criterion is used, and the crack size for failure, a_f, is a function of the fracture toughness and the maximum applied stress. (ii) Fatigue life is a strong function of the fatigue crack growth kinetics and of geometry, the influence of applied stress being a strong function of the form of rate equation in (7.10). (iii) Fatigue life is also affected strongly by the initial crack size, and less so by the final crack size. For $n = 2$, for example, doubling a_i reduces the fatigue life by more than a factor of 2. If a_i is much smaller

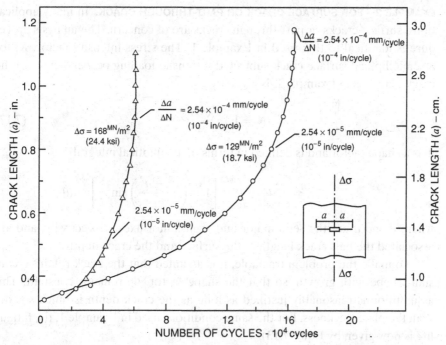

Figure 7.9. Constant load-amplitude fatigue crack growth curves for a mill-annealed Ti-6Al-4V alloy tested in vacuum at room temperature [4].

than a_f, the final crack size (based on fracture toughness) would have a negligible effect on fatigue life. The effect of these variables may be readily seen by examining actual fatigue crack growth data on some titanium alloys (Figs. 7.9 and 7.10) [4].

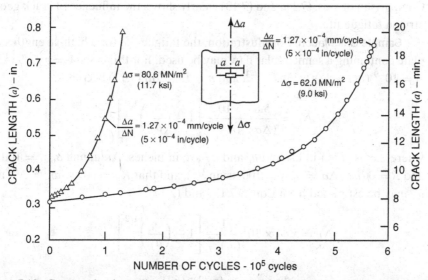

Figure 7.10. Constant load-amplitude fatigue crack growth curves for a mill-annealed Ti-6Al-4V alloy tested in dehumidified argon at 140C [4].

EXAMPLE 2 – FOR SURFACE CRACK OR PART-THROUGH CRACK. In many applications, surface cracks or part-through cracks are of concern. The analysis procedure is identical to that used in Example 1. The stress intensity factor, K, for a semielliptical surface crack subjected to tensile loading perpendicular to the crack plane, as in Example 1, is given by [4]:

$$K = 1.1\sigma \frac{\sqrt{\pi a}}{\Phi} \tag{7.12}$$

Φ is a shape factor and is defined in terms of an elliptical integral

$$\Phi = \int_0^{\pi/2} \left[1 - \left(\frac{c^2 - a^2}{c^2} \right) \sin^2 \theta \right]^{1/2} d\theta$$

where c and a are the semi-major and semi-minor axes, respectively, and are associated the half-crack length at the surface and the crack depth.

To make the problem tractable, it is assumed that the crack retains a constant shape with growth, so that the shape factor, Φ, remains constant. This assumption is reasonably justified as long as the crack depth is much smaller than the plate thickness. For the same conditions used in Example 1, the fatigue life is now given by Eqn. (7.13).

$$N_F = \frac{\Phi^2}{(1.1)^{2n} \pi^n A (n-1)(\Delta\sigma)^{2n} (a_i/\Phi^2)^{(n-1)}} \left[1 - \left(\frac{a_i}{a_f} \right)^{(n-1)} \right] \quad (n > 1) \tag{7.13}$$

Assuming failure to occur in plane strain, then

$$a_f = \frac{\Phi^2 K_{Ic}^2}{1.21\pi \sigma_{\max}^2} \tag{7.14}$$

Comparison of Eqs. (7.11) and (7.13) clearly shows the influence of crack geometry on fatigue life.

Sample Calculation. For illustration, the fatigue life for a high-strength steel plate containing a semicircular flaw may be used. For this case, $\Phi = \pi/2$. Taking $A = 10^{-9}$ (in./cycle)(ksi/in.$^{1/2}$)$^{-3}$, and $n = 1.5$, Eqn (7.13) becomes,

$$N_F = \frac{5.84 \times 10^9}{(\Delta\sigma)^3 (a_i)^{1/2}} \left[1 - \left(\frac{a_i}{a_f} \right)^{1/2} \right]$$

where $\Delta\sigma$ is given in ksi and a_i and a_f are in inches. Assuming $\sigma_{\max} = 100$ ksi and $R = 0$ (*i.e.*, $\Delta\sigma = \sigma_{\max} = 100$ ksi/in.$^{1/2}$), and that $K_{Ic} = 60$ ksi-in.$^{1/2}$, N_F, and a_f may be estimated from Eqns. (7.13) and (7.14).

$$N_F = 5.84 \times 10^3 \frac{1}{(a_i)^{1/2}} \left[1 - \left(\frac{a_i}{a_f} \right)^{1/2} \right]$$

$$a_f = \frac{\Phi^2 K_{Ic}^2}{1.21\pi \sigma_{\max}^2} = \frac{(\pi/2)^2 (60)^2}{1.21\pi (150)^2} = 0.233 \text{ in.}$$

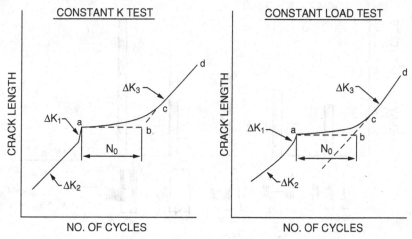

Figure 7.11. Schematic illustration of delay in fatigue crack growth and definition of delay, N_D [4].

If NDI techniques can detect initial cracks as small as 10^{-2} in., then using $a_i = 10^{-2}$ in., the estimated resulting fatigue life would be about 5×10^4 cycles. If, on the other hand, the initial crack size must be estimated from proof testing, say at 150 ksi, the assumed initial crack size for use in the fatigue analysis must be equal to the critical size for the proof stress (σ_{pr}), Eqn. (7.14).

$$a_i = \frac{\Phi^2 K_{Ic}^2}{1.21\pi\sigma_{pr}^2} = \frac{(\pi/2)^2(60)^2}{1.21\pi(150)^2} = 0.104 \, \text{in.}$$

The service life is then reduced to about 6×10^3 cycles. This example illustrates the importance of avoiding, or reducing the size of initial defects in the structure, and the need for improved methods for nondestructive inspection.

VARIABLE AMPLITUDE LOADING. Many attempts have been made to predict fatigue lives under variable amplitude loading. It is recognized that crack growth rates can be significantly affected by load interactions. These interactions can produce "acceleration" or "retardation or delay" in crack growth, as illustrated schematically in Fig. 7.11, or in delay as shown in Fig. 7.12. For randomized spectra, it appears that the rate of crack growth may be characterized reasonably well using the root-mean-squared (rms) value of ΔK (ΔK_{rms}). For more ordered spectra, on the other hand, numerical integration approaches (such as, AFGROW [6] and FASTRAN [7]) are used. These codes require extensive supporting data for each material, and contain adjustable parameters that are not always transparent to the users. As such, experimental validations are essential.

7.4 Effects of Loading and Environmental Variables

It has been shown that fatigue life is influenced by the fatigue crack growth kinetics and is reflected through changes in A and n in the power-law representation, for

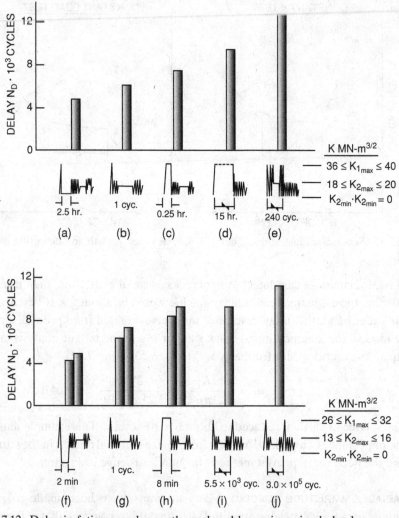

Figure 7.12. Delay in fatigue crack growth produced by various simple load sequences for mill-annealed Ti-6Al-4V alloy tested in air at room temperature [4].

example, in Eqn. (7.10). The influence of various loading and environmental variables on fatigue life may be examined, therefore, in terms of their effects on the kinetics of fatigue crack growth.

Fatigue crack growth is affected by a range of metallurgical, environmental, and mechanical (loading) variables. They have been extensively reviewed [4]. It is important to note that crack growth is influenced by a broad range of loading variables (*e.g.*, maximum load, load ratio (minimum/maximum), frequency, waveform, etc.), some of which can interact with the environment. Many of the observed effects of loading variables can be traced directly to environmental interactions, and will be considered in detail in the following chapters. On the basis of data that have been gathered over the past twenty-plus years, the response of fatigue crack growth may be grouped into three basic types and be discussed in relation to K_{Iscc}, the

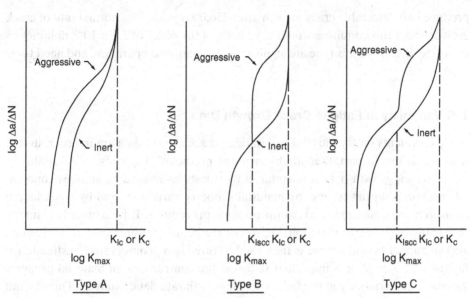

Figure 7.13. Types of corrosion (environmentally affected) fatigue crack growth response [4].

threshold stress intensity factor for stress corrosion cracking (Fig. 7.13) [4]. Type A behavior is typified by the aluminum-water system. The observed environmental effects result from the interaction of fatigue and environmental attack by hydrogen that is released by the water-metal reaction. Type B behavior is represented by the hydrogen steel system. Environment-enhanced crack growth is directly related to sustained-load crack growth (during the higher-load portion of the fatigue cycle), with no interaction effects. Type C represents the behavior of most alloy-environment systems. Above K_{Iscc}, the behavior approaches that of type B, whereas, below K_{Iscc}, the behavior tends toward type A, with the associated interaction effects.

The effects of load interactions on fatigue crack growth under variable amplitude loading can be very large. Acceleration in the rate of fatigue crack growth may be encountered with increases in cyclic-load amplitude, and delay in fatigue crack growth is associated with decrease in load amplitude. Acceleration in fatigue crack growth is generally significant at high K levels (or at high loads). Since high loads are expected to occur infrequently and are usually of short duration, the influence of crack acceleration on fatigue life, typically, may be neglected. Delay, N_D (or the retardation in the rate of fatigue crack growth), on the other hand, can be particularly large at the lower (normal operating) K levels and needs to be taken into consideration in efficient design. Delay is affected by a broad range of loading and environmental variables. The influences of some loading variables were shown previously in Figs. 7.11 and 7.12. The significance of these effects may be illustrated by the following example on an annealed Ti-6Al-4V alloy. Loading at a K of about 11 MPa-m$^{1/2}$ with $R = 0$, following a single high-load excursion to 30.8 MPa-m$^{1/2}$, produced no detectable crack growth after 450,000 cycles. The normal rate of crack growth under this condition would have been of the order of 2.5×10^{-5} millimeters per cycle. Effects of this type are significant to design and operation, and need to be recognized.

7.5 Variability in Fatigue Crack Growth Data

Some comments on the variability in fatigue crack growth data are in order, in that it would define, in part, the variability in the predicted fatigue lives. Variability in fatigue crack growth data is introduced by variations in loading and environmental conditions during testing, by material property variations, and by crack length measurement techniques and data-processing procedures. It is commonly assumed that material property variations represent the primary source of data variability, and an "upper bound" curve is then used to provide a "conservative" estimate for fatigue life. Indeed, it is important to assess the contribution of material property variations to variability in the fatigue crack growth rate data. Study by Thomas and Wei [9] has shown, however, that a significant portion of the variability may have come from the experimental and data analysis procedures. The use of this information on "variability" in design, therefore, could yield misleading results and should be viewed with caution.

7.6 Summary Comments

In this chapter, the fracture mechanics approach to stress corrosion and fatigue crack growth in design is described from a phenomenological perspective. This approach assumes the pre-existence of cracks, or cracklike entities, in a structural component, and focuses attention on their growth and the resulting impact on strength and life (namely, structural integrity and durability). Successful application of this approach to design and system management depends on the ability to identify and cope with problems in stress/structural analysis, in defining the service loading and environmental conditions, in properly characterizing the kinetics of crack growth under these conditions, and in synthesizing all of this information in the durability and reliability analyses.

Heretofore, fracture mechanics has been used in characterizing the driving force for crack growth, and in experimental measurements (often without careful control of the chemical and thermal environments) to define material response. It was then used for structural integrity and durability analyses. It was now recognized that the response of different materials to different chemical environments and at different temperatures can be very different. In the following chapters, the influences of chemical, thermal, and microstructural variables on crack growth response will be explored. The probabilistic impact of these variables on durability and structural integrity will be explored. Although the author has devoted most of his professional life to studies in this area, much more needs to be done to broaden the scope of understanding over the wide range of materials and environments that are encountered in practice.

REFERENCES

[1] Brown, B. F. and Beachem, C. D., "A Study of the Stress Factor in Corrosion Cracking by use of the Pre-cracked Cantilever Beam Specimen," Corrosion Science, 5 (1965), 745–750.

[2] Wei, R. P., "Application of Fracture Mechanics to Stress Corrosion Cracking Studies," in Fundamental Aspects of Stress Corrosion Cracking, NACE, Houston, TX (1969), 104.

[3] Wei, R. P., Novak, S. R., and Williams, D. P., "Some Important Considerations in the Development of Stress Corrosion Cracking Test Methods," AGARD Conf. Proc. No. 98, Specialists Meeting on Stress Corrosion Testing Methods, 1971, Materials Research and Standards, ASTM, 12, 9 (1972), 25.

[4] Wei, R. P., "Fracture Mechanics Approach to Fatigue Analysis in Design," J. Eng'g. Mat'l. & Tech., 100 (1978), 113–120.

[5] Williams, D. P., and Nelson, H. G., "Gaseous Hydrogen – Induced Cracking of Ti-5Al-2.5Sn," Met. Trans., 3, 8 (1972), 2107.

[6] Chu, H. C., and Wei, R. P., "Stress Corrosion Cracking of High-Strength Steels in Aqueous Environments," Corrosion, A6, 6 (1990), 468–476.

[7] Harter, J., ARGROW Program; AFGROW/VASM, http://www.stormingmedia. us/13/1340/A134073.html, (2004).

[8] Newman, J. C., Jr., "FASTRAN II – A fatigue crack growth structural analysis program," NASA TM-104159 (1992).

[9] Thomas, J. P., and Wei, R. P., "Standard-Error Estimates for Rates of Change from Indirect Measurements", Technometrics, 38,1 (1996), 59–68.

8 Subcritical Crack Growth: Environmentally Enhanced Crack Growth under Sustained Loads (or Stress Corrosion Cracking)

8.1 Overview

In Chapter 7, the subject of subcritical crack growth (namely, stress corrosion cracking and corrosion fatigue) was treated from a phenomenological perspective. The emphasis, by and large, is focused on the development of design data to cover a limited range of service conditions, rather than a broad-based understanding. As such, the influences of material composition and microstructure, and their interactions with the external chemical and thermal environment (*e.g.*, atmospheric moisture and sea water) are not fully addressed. As such, the data are only of limited value, and cannot be "extrapolated" to cover other loading and environmental conditions.

In this and the next chapter, the contributions to the understanding and modeling of the effects of conjoint actions of loading, and chemical and thermal variables on a material's crack growth response are highlighted. The presentation draws principally on results from the author's laboratory. For clarity, the influences of gaseous and aqueous environments under sustained or statically applied loads (or stress corrosion cracking) are considered here. Those for fatigue crack growth are highlighted in Chapter 9. Illustrations (to a large extent constrained by the "windows of opportunity") are drawn from research in the author's laboratory, and will cover high-strength steels in gaseous and aqueous environments, nickel-base superalloys in oxygen, and ceramics in water. For fatigue crack growth, the materials include aluminum and titanium alloys and steels. Understanding is derived through coordinated experiments and analyses that probe the underlying chemical, mechanical, and materials interactions for crack growth. The more extensive treatment of nickel-base superalloys highlights oxygen (perhaps others) as a potential "embrittler" and the limited study on ceramics serves to broaden the perspective on environmentally enhanced crack growth.

Modeling of crack growth response, to reflect control of crack growth by the transport of deleterious species to the crack tip, or surface/electrochemical reaction of the newly created surfaces at the crack tip, or diffusion of hydrogen/oxygen atoms/ions ahead of the crack tip, are presented here and in Chapter 9 to reflect differences in loading and crack growth response. Specifically, under sustained loading

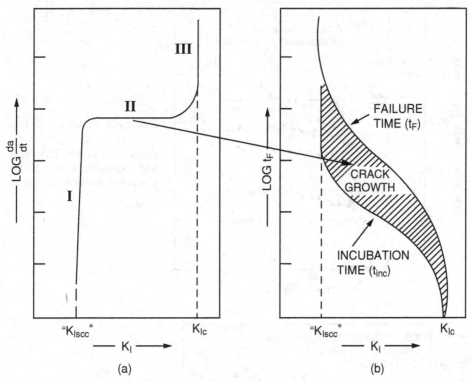

Figure 8.1. Schematic illustration of the growth rate (kinetics) and failure time (life) versus driving force responses under sustained (constant) loads [1].

(*i.e.*, for stress corrosion cracking (SCC)), crack growth is "continuous" and allows the surface reactions to be completed. As such, crack growth rate is inversely proportional to the amount of time to complete the surface reaction. The sustained-load crack growth rates reflect directly the underlying rate-controlling process (namely, environment transport, surface reaction, or diffusion of the damaging specie). For fatigue crack growth (FCG), on the other hand, the extent of reaction is limited by the cyclic period of loading. As such, the environmental response is reflected through the crack growth rate dependence on the inverse, or the inverse square root, of loading frequency. This difference is highlighted through the parallel, but different, "environmental" formulation in this and the following chapters.

8.2 Phenomenology, a Clue, and Methodology

Well-controlled experimental studies on a range of materials, in simple chemical environments, have shown a typical response as depicted schematically in Fig. 8.1 [1]. Crack growth begins from some threshold stress intensity factor level (K_{th} or K_{Iscc}). The growth rate then rises rapidly, reaching a plateau and then increases again rapidly to reach failure at the material's fracture toughness value of K_{Ic} or K_c. Figures 8.2, 8.3, and 8.4 show data on a titanium alloy in hydrogen as a function of temperature [2], and on an AISI 4340 steel in 0.6 N NaCl solution as a function of temperature, and on an AISI 4130 steel as a function of solution chemistry at

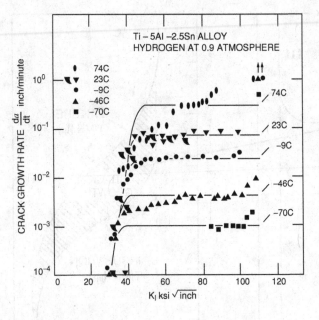

Figure 8.2. Influence of temperature on crack growth response for a Ti-5Al-2.5Sn alloy under sustained load in hydrogen at 0.9 atmosphere [2].

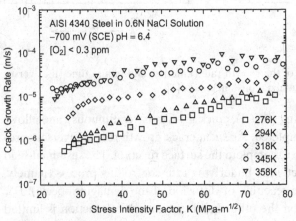

Figure 8.3. Influence of temperature on crack growth response for an AISI 4340 steel under sustained load in a 0.6 N NaCl solution (pH = 6.4) at 276 to 358 K [3].

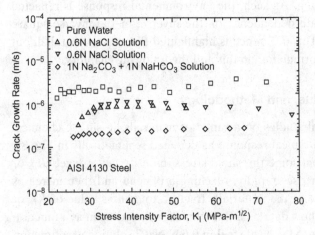

Figure 8.4. Influence of solution chemistry on crack growth response for an AISI 4130 steel under sustained load [3].

room temperature [3]. The indicated near-independence of crack growth rate on the mechanical crack-driving force K, and its dependence on thermal and chemical environments, provide a link for examining the process(s)/mechanism(s) that control crack growth, and for the development of tools for their mitigation and for design.

The essential methodology for understanding and developing effective tools for design and sustainment of engineered systems involves the development of understanding of the damage evolution processes and of tools for their mitigation and control. It involves the use of well designed experiments to probe the underlying mechanisms and rate-controlling processes for crack growth through:

- influence of temperature, frequency, etc.
- partial pressure and gaseous species (for gaseous environments)
- ionic species, concentration, pH, etc. (for aqueous environments)
- supporting microstructural and chemical investigations.

8.3 Processes that Control Crack Growth

The processes that are involved in the enhancement of crack growth in high-strength alloys by gaseous environments (*e.g.*, hydrogen and hydrogenous gases (such as H_2O and H_2S), or oxygen), are illustrated schematically in Fig. 8.5 and are as follows [1]:

1. Transport of the gas or gases to the crack tip.
2. Reactions of the gas or gases with newly produced crack surfaces to evolve hydrogen, or surface oxygen (namely, physical and chemical adsorption).
3. Hydrogen or oxygen entry (or absorption).
4. Diffusion of hydrogen or oxygen to the fracture (or embrittlement) sites.
5. Partition of hydrogen or oxygen among the various microstructural sites.
6. Hydrogen-metal or oxygen-metal interactions leading to embrittlement (i.e., the embrittlement reaction) at the fracture site.

Figure 8.5. Schematic diagram of processes involved in the enhancement of crack growth in gaseous environments [1].

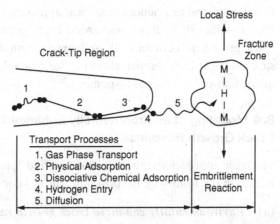

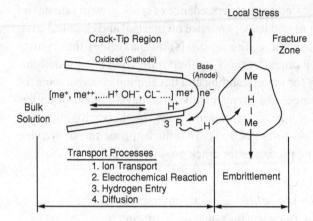

Figure 8.6. Schematic diagram of processes involved in the enhancement of crack growth in aqueous environments [1].

For crack growth in aqueous solutions, the corresponding processes are as follows, and are schematically shown in Fig. 8.6 [1]:

1. Liquid phase transport along the crack

    • Convection (pressure gradient)
    • Diffusion (concentration gradient)
    • Electromigration (potential gradient)

2. Coupled electrochemical reaction at the crack tip/dissolution (go no further for dissolution mechanism).
3. Hydrogen entry (or absorption).
4. Diffusion and partitioning of hydrogen to the fracture (or embrittlement) sites.
5. Hydrogen-metal interactions leading to embrittlement (i.e., the embrittlement reaction) at the fracture site. (Although metal dissolution has been considered as the mechanism for stress corrosion crack growth, fractographic evidence to date does not support it as a viable mechanism.)

The various processes, and their inter-relationships, are depicted in the schematic diagrams in Fig. 8.7. Figure 8.8, on the other hand, represents the more traditional empirical or phenomenological approach in which empiricism resides. More recent studies of hydrogen-enhanced crack growth in steels [3], moisture-induced crack growth in ceramics [4], and oxygen-enhanced crack growth in nickel-base superalloys at high temperatures [5], for example, broaden the understanding and quantification of material response.

8.4 Modeling of Environmentally Enhanced (Sustained-Load) Crack Growth Response

The basic approach to the understanding and "prediction" of crack growth response resides in the following: (a) postulate and (b) corollary, *i.e.*,

(a) *"Environmentally enhanced crack growth results from a <u>sequence</u> of processes and is <u>controlled</u> by the <u>slowest</u> process in the sequence."*

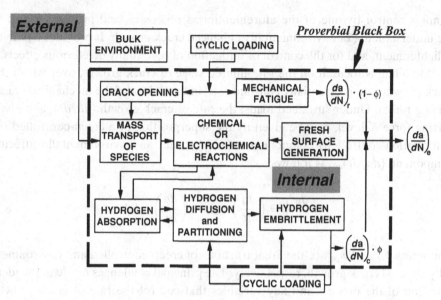

Figure 8.7. Block diagram showing the various processes that are involved, and their relationships, in the environmental enhancement of crack growth.

(b) *Crack growth response reflects the dependence of the rate-controlling process **on the environmental, microstructural, and loading variables.***

This fundamental hypothesis reflects the existence of a region (*i.e.*, stage II) in crack growth response over which the growth rate is essentially constant (*i.e.*, independent of the mechanical crack-driving force). The existence of this rate-limited region

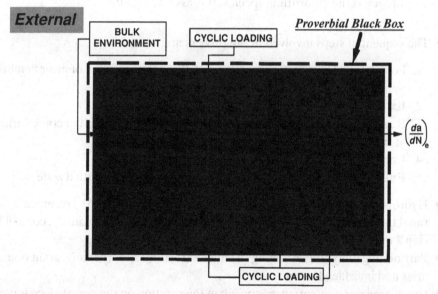

Figure 8.8. Illustration of a more empirical approach in which the controlling processes (Fig. 8.7) are by-and-large hidden.

signifies control by one of the aforementioned processes, and provides a link to the understanding of environmentally enhanced crack growth. It provides a path to enlightenment, and for the control or mitigation of potentially deleterious effects.

Modeling is focused on the rate-limited stage of crack growth, over which the crack growth rate is essentially constant (*i.e.*, independent of the mechanical crack-driving force). Under sustained loads, the rate of crack growth (da/dt), at a given driving force K level, may be given by the superposition of a creep-controlled (or deformation-controlled) component, $(da/dt)_{cr}$, and an environmentally affected component, $(da/dt)_{en}$, as follows:

$$\left(\frac{da}{dt}\right) \cong \left(\frac{da}{dt}\right)_{cr} \phi_{cr} + \left(\frac{da}{dt}\right)_{en} \phi_{en} \qquad (8.1)$$

The terms ϕ_{cr} and ϕ_{en} are the areal fractions of creep-controlled and environmentally affected crack growth, respectively. The principal challenges reside in the identification of the process and key variables that control the rate of crack growth, and in the quantification and modeling of the influences of these variables on crack growth response in terms of these key variables. The crack growth rate is governed by the crack-driving force given by the stress intensity factor K_I, and reflects "control" (*i.e.*, rate limited) by the underlying "deformation and chemical" processes. The overall modeling is treated as a pseudo-static problem, and is viewed incrementally.

Modeling Assumptions

The modeling was first developed for crack growth in gaseous environments in which hydrogen is the embrittling species. It is assumed that:

- The sequential steps involved in the process are:
  1. Formation of new surfaces; *i.e.*, growth through the region of prior "embrittlement"
  2. External transport of gas to the (new) crack tip
  3. Reaction (dissociative chemisorption) with the newly created crack surface at the crack tip to produce hydrogen
  4. Entry/diffusion of hydrogen to the embrittlement zone
  5. Embrittlement reaction, or re-establishment of an embrittled zone

- Hydrogen entry/diffusion and embrittlement (steps 4 and 5) are much more rapid than gas transport and surface reaction (steps 2 and 3); namely, control by step 2 or step 3.
- Partitioning of hydrogen among the microstructural sites (namely, grain boundaries and interfacial sites).
- Crack grows or new surfaces form when the reaction on the new surface is complete (for the sustained-load case here); namely, when θ approaches 1.0.

Figure 8.9. Schematic representation of the transport of gases along a crack to its tip.

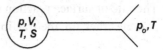

8.4.1 Gaseous Environments

Studies of environment-enhanced crack growth in gaseous environments have shown that crack growth may be controlled by (i) the rate of transport of the environment (along the crack) to the crack tip, (ii) the rate of surface reactions with the newly created crack surfaces to evolve hydrogen, or (iii) the rate of diffusion of hydrogen into the "process zone" ahead of the crack tip. In this simplified, chemical-based model, the competition between transport and surface reaction is considered.

For the simplified model, the crack-tip region is considered to be a closed volume V that is connected to the external environment through a narrow "pipe" (the crack) (see Fig. 8.9). The crack-tip region is characterized by the pressure (p), its volume (V), and surface (S), and the temperature (T), and by the number of gas molecules (n) that are present. The environment at the crack mouth is characterized by the external gas pressure (p_o) and temperature (T). The two temperatures are assumed to be equal. These quantities are related through the perfect gas law as follows:

$$pV = nkT \qquad (8.2)$$

where n is the number of gas molecules in the crack-tip volume, and k is Boltzmann's constant. Treating the crack-tip region as a constant volume system, the rate of change in pressure is related to the rate of change in the number of gas molecules in the region; namely:

$$\frac{dp}{dt} = \frac{kT}{V}\frac{dn}{dt} \qquad (8.3)$$

where the rate of change in the number of molecules in the gas phase, in the crack-tip volume, is related to the rate of consumption, by reactions with the cavity wall, and the rate of supply, by ingress along the crack; namely:

$$\text{Consumption:} \quad \frac{dn}{dt} = -SN_o\frac{d\theta}{dt}$$

$$\qquad (8.4)$$

$$\text{Supply:} \quad \frac{dn}{dt} = \frac{F}{kT}(p_o - p)$$

where n = number of gas molecules in the crack-tip "cavity"; S = surface area of the "cavity"; N_o = density of metal atoms on the surface; θ = fractional surface coverage or atoms that have reacted; F = volumetric flow rate coefficient; k = Boltzmann's constant; and p_o and p = the pressure outside and at the crack tip, respectively.

By inserting Eqn. (8.4) into Eqn. (8.3), conservation of mass yields the rate of change in pressure at the the crack tip, or the conservation of mass in terms of pressure as:

$$\frac{dp}{dt} = -\frac{SN_okT}{V}\frac{d\theta}{dt} + \frac{F}{V}(p_o - p) \qquad (8.5)$$

The rate of surface reaction is given in terms of a reaction rate constant k_c, pressure p at the crack tip, the fraction of open (unreacted) sites $(1 - \theta)$:

$$\frac{d\theta}{dt} = k_c p f(\theta) = k_c p (1 - \theta) \tag{8.6}$$

which assumes first-order reaction kinetics. Combining Eqns. (8.5) and (8.6) and solving for p, one obtains:

$$\frac{dp}{dt} = -\frac{SN_o kT}{V} k_c p (1 - \theta) + \frac{F}{V} (p_o - p)$$

$$p = \frac{p_o - \frac{V}{F}\frac{dp}{dt}}{\frac{SN_o kT}{F} k_c p (1 - \theta) + 1} \tag{8.7}$$

As a steady-state approximation, it is assumed that $dp/dt = 0$. The pressure at the crack tip then becomes:

$$p = \frac{p_o}{\frac{SN_o kT}{F} k_c p (1 - \theta) + 1} \tag{8.8}$$

Examination of Eqn. (8.8) shows that there are two limiting cases: (i) For $k_c \gg 1$, $p \ll p_o$ whereby the reaction would be limited by the rate of transport of the deleterious gas to the crack tip. (ii) For $k_c \ll 1$, the pressure p at the crack tip is approximately equal to the external pressure p_o whereby the reaction would be limited by the rate of reaction of the deleterious gas with the crack-tip surfaces.

Substituting Eqn. (8.8) into Eqn. (8.6) for surface reaction, one obtains:

$$\frac{d\theta}{dt} = k_c p (1 - \theta) = \frac{k_c p_o (1 - \theta)}{\frac{SN_o kT}{F} k_c (1 - \theta) + 1} \tag{8.9}$$

By separating the variables θ and t, Eqn. (8.9) becomes:

$$\left[\frac{SN_o kT}{F} k_c + \frac{1}{(1 - \theta)} \right] d\theta = k_c p_o dt$$

By integration, one obtains the following relationship for the fractional surface coverage θ, or the extent of surface reaction, as a function of time, namely,

$$\frac{SN_o kT}{F} k_c \theta - \ell n (1 - \theta) = k_c p_o t \tag{8.10}$$

The solution, Eqn. (8.10), yields two limiting cases: (a) when the gas-metal reactions are very active (i.e., when k_c is very high), the production of "embrittling" species is governed by the rate of its transport to the crack tip, and (b) when the surface reaction rates are slow, crack growth is controlled by the rate of these reactions to evolve hydrogen. Namely:

$$\text{Transport control:} \quad \theta \approx \frac{F p_o}{S N_o k T} t$$

(8.11)

$$\text{Surface reaction control:} \quad \theta \approx 1 - \exp(-k_c p_o t)$$

The rate of environmentally enhanced crack growth is essentially inversely proportional to the time required to cover (or for the environment to react) with an increment of newly exposed crack surface. It is estimated based on the time required for the environment to fully react with an increment of newly produced crack surface, or in terms of the rate of supply of the environment and the rate of consumption (surface reaction); namely, mass balance.

8.4.1.1 Transport-Controlled Crack Growth

For transport-controlled crack growth, the functional dependence of crack growth rate is simply determined from the conservation of mass, in which the rate of consumption of gas molecules through reactions with the newly created metal surfaces by cracking is governed by the rate of supply of the deleterious gas species along the crack. In other words, the newly created crack surface is so active that every gas molecule that arrives at the crack tip is assumed to react "instantly" with it. The transport of gas along the crack is modeled in terms of Knudsen (molecular) flow [8], with drift velocity V_a and the crack modeled as a narrow capillary of height δ, width (representing the thickness of the specimen/plate) B, and length L, and is given by Eqn. (8.12):

$$F = \frac{4}{3} V_a \frac{\delta^2 B}{L}$$

(8.12)

where

$$V_a = \left(\frac{8kT}{\pi m} \right)^{1/2}; \qquad m = \frac{M}{N_a}$$

$$k = \text{Boltzmann's constant}$$
$$m = \text{mass of a gas molecule}$$
$$M = \text{gram molecular weight of the gas}$$
$$N_a = \text{Avogadro's Number}$$

Substituting the mass of the gas molecule, in terms of its gram molecular weight, and Avogadro's number, V_a and F are given as follows:

$$V_a = \left(\frac{8 N_a k T}{\pi M} \right)^{1/2} = 1.45 \times 10^2 \left(\frac{T}{M} \right)^{1/2} \quad \text{m/s}$$

(8.13)

$$F = \frac{4}{3} V_a \frac{\delta^2 B}{2L} = 97 \frac{\delta^2 B}{2L} \left(\frac{T}{M} \right)^{1/2} \quad \text{m}^3/\text{s}$$

The functional dependence for transport-controlled crack growth is obtained simplify by equating the rate of consumption of the gas by reactions with the newly
created crack surface and the rate of supply of gas by Knudsen flow along the crack.
The rate consumption is equal to the rate at which new crack surface sites (atoms)
are created, and is given by:

$$N_o \alpha (2B) \frac{da}{dt} \quad \text{(number of surface sites created per unit time)}$$

where N_o is the density of surface sites, B is the thickness of the material, da/dt
is the crack growth rate, and α (greater than 1) represents a roughness factor that
increases the effective surface area. The rate of supply of gas through the crack, in
atomic units, is given by:

$$\frac{F}{kT}(p_o - p)$$

Equating the rates of supply and consumption leads to:

$$N_o \alpha (2B) \frac{da}{dt} (p_o - p) \approx \frac{F p_o}{kT}; \quad \text{because } p_o \gg p$$

Because, as seen previously,

$$F = \frac{4}{3} V_a \frac{\delta^2 B}{2L}; \quad V_a = \left(\frac{8 N_a k T}{\pi M} \right)^{1/2} \propto T^{1/2}$$

Therefore,

$$\frac{da}{dt} \propto \frac{p_o}{T^{1/2}} \quad \text{Transport control} \tag{8.14}$$

8.4.1.2 Surface Reaction and Diffusion-Controlled Crack Growth

If the rate of transport of gases along the crack were sufficiently fast, then crack
growth would be controlled (rate limited) by the rate of surface reactions with the
newly created crack surface. Assuming, for simplicity, that the reactions follow first-
order kinetics, the rate of increase in the fractional surface coverage θ is given by
Eqn. (8.15):

$$\frac{d\theta}{dt} = k_c p_o (1 - \theta); \quad k_c = k_{co} \exp\left(-\frac{E_S}{RT} \right) \tag{8.15}$$

where k_c is the reaction rate constant that reflects a thermally activated process
represented by a rate constant k_{co} and an activation energy E_S. Equation (8.15) may
be integrated to yield the surface coverage θ as a function of time or the time interval
Δt_c to reach a "critical" coverage θ_c (say, 0.9 or 0.95); i.e.:

$$\theta = 1 - \exp(-k_c p_o t)$$
or
$$\Delta t_c = \int_0^{\Delta t_c} dt = \frac{1}{k_c p_o} \int_0^{\theta_c} \frac{d\theta}{1 - \theta} = \frac{1}{k_c p_o} \ln(1 - \theta_c)$$

The functional dependence of crack growth rate on pressure and temperature (for a monotonic gas) is deduced from the foregoing relationship as follows:

$$\frac{da}{dt} \approx \frac{\Delta a}{\Delta t} \propto \frac{1}{\Delta t_c} \propto k_c p_o \Rightarrow \frac{da}{dt} \propto p_o \, \exp\left(-\frac{E_S}{RT}\right) \qquad (8.16)$$

More generally, for diatomic gases, such as hydrogen, the following form for surface reaction control is used:

$$\frac{da}{dt} \propto p_o^m \, \exp\left(-\frac{E_S}{RT}\right) \qquad (8.17)$$

If the transport and surface reaction processes are rapid (*i.e.*, not rate limiting), then crack growth would be controlled by the rate of diffusion of the embrittling species into the fracture process zone ahead of the crack tip. For diffusion-controlled crack growth, therefore, the rate equation assumes the following form:

$$\frac{da}{dt} \propto p_o^m \, \exp\left(-\frac{E_D}{2RT}\right) \qquad (8.18)$$

The exponent m in Eqns. 8.17 and 8.18 is typically assumed to be equal to 1/2 for diatomic gases, such as hydrogen; but the number m is used here to recognize the possible existence of intermediate states in the dissociation from their molecular to atomic form. The factor of 2 in the exponential term gives recognition for the dissociation of diatomic gases, such as hydrogen (H_2).

8.4.2 Aqueous Environments

Cracking problems in aqueous environments, or stress corrosion cracking (SCC), has been the traditional domain of corrosion chemists. The prevailing view before the 1980s was that SCC is the result of stress-enhanced dissolution of material at the crack tip. This view was supported by potentiostatically controlled, transient ("straining" and "scratching") electrode experiments that suggested very rapid dissolution of the freshly exposed surface was supported by very high transient currents shown by these experiments. Beginning in the 1970s, there was growing concern with respect to the interpretation and applicability of these findings. It was suspected that the use of a potentiostat might have adversely affected the "repassivation current" measurements.[1]

A series of experiments were conducted at Lehigh University, in which the repassivation currents were measured by *in situ* fracture of notched round specimens under open-circuit conditions (*i.e.*, without potentiostatic control); see, for example, Figs. 8.10 and 8.11. These results were more consistent with the repassivation of a freshly exposed surface. Taking the inverse of the time to reach a given

---

[1] Demonstrated by the recognition that the maintenance of "a constant potential" required the potentiostat to send a "large" current through the counter-electrode, which was superimposed on to, and misinterpreted as the repassivation current.

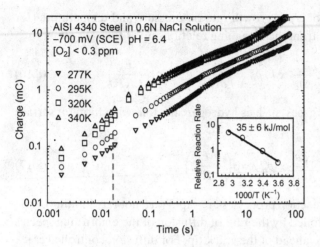

Figure 8.10. Charge transfer versus time and temperature for the reactions of bare surfaces of AISI 4340 steel with 0.6 N NaCl solution at −700 mV (SCE), pH = 6.4 [8, 9].

charge level (say 0.2 mC) as a rate of reactions, the estimated activation energy for the reactions is depicted in the insect to the figures.

Analogous to surface reaction-controlled crack growth in gaseous environments, electrochemical reaction-controlled crack growth is given by:

$$\left(\frac{da}{dt}\right)_{II} \approx \frac{\Delta a}{\Delta t} \approx \frac{\Delta a}{\frac{1}{k}\int_0^{\theta_c}\frac{d\theta}{1-\theta}d\theta} \propto k \propto k_o \exp\left(-\frac{E_S}{RT}\right)$$

or   (8.19)

$$\left(\frac{da}{dt}\right)_{II} = C_S \exp\left(-\frac{E_S}{RT}\right)$$

The temperature dependence reflects the activation energy for electrochemical reactions with bare surfaces. The crack growth rate reflects the dependence on anion type, concentration, and temperature.

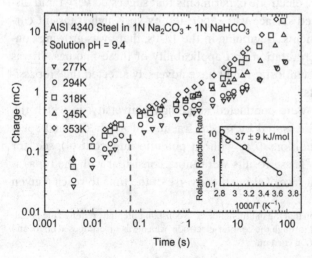

Figure 8.11. Charge transfer versus time and temperature for the reaction of bare surfaces of AISI 4340 steel with 1 N Na_2CO_3 + 1 N $NaHCO_3$ solution, pH = 9.4 [8, 9].

If the electrochemical reactions are rapid, control may be passed on to the diffusion of damaging species (hydrogen) into the region ahead of the growing crack tip. As such, the model for diffusion-controlled crack growth may be applied directly.

8.4.3 Summary Comments

The foregoing models provide the essential link between the fracture mechanics and surface chemistry/electrochemistry aspects of crack growth response. Crack growth response, in fact, is the response of a material's microstructure to the conjoint actions of the mechanical and chemical driving forces. In the following sections, the responses in gaseous and aqueous environments are illustrated through selected examples from the work of the author and his colleagues (faculty, researchers, and graduate students).

8.5 Hydrogen-Enhanced Crack Growth: Rate-Controlling Processes and Hydrogen Partitioning

Crack growth, under sustained loading, in high-strength steels exposed to gaseous and aqueous environments has been widely studied from a multidisciplinary point of view. A series of parallel fracture mechanics and surface chemistry studies on high-strength steels, exposed to hydrogen-containing gases (such as, hydrogen, hydrogen sulfide, and water vapor) and to aqueous electrolytes, has provided a clearer understanding of hydrogen-enhanced crack growth [3]. It is now clear that hydrogen-enhanced crack growth is controlled by a number of processes in the embrittlement sequence (see Fig. 8.5); namely, (i) transport of the gas or gases, or electrolyte, to the crack tip; (ii) the reactions of the gases/electrolytes with newly formed crack surfaces to evolve hydrogen (namely, physical and dissociative chemical adsorption in sequence); (iii) hydrogen entry (or absorption); (iv) diffusion of hydrogen to the fracture (or embrittlement) sites; and (v) hydrogen-metal interactions leading to embrittlement (*i.e.*, the embrittlement sequence, or cracking). Modeling of crack growth response must be appropriate to the rate-controlling process and reflect the appropriate chemical, microstructural, environmental, and loading variables.

For modeling, attention has been focused on stage II of sustained load crack growth, where the crack growth rate reflects the underlying rate-controlling process, and is essentially independent of the mechanical driving force. The modeling effort was guided by extensive experimental observations (see [3]). The stage II crack growth responses for an AISI 4340 steel, in hydrogen, hydrogen sulfide, and water, at different temperatures are shown in Fig. 8.12, along with identification of the rate-controlling processes. In the low-temperature region, below about 60°C, cracking followed the prior-austenite grain boundaries, with a small amount of quasi-cleavage that reflected cracking along the martensite lath or patch boundaries, and the $\{110\}_{\alpha'}$ and $\{112\}_{\alpha'}$ planes through the martensites (see Fig. 8.13). Cracking became dominated by the microvoid coalescence mode of separation as the temperature increased into the region above about 80°C. Suitable models, therefore, had to

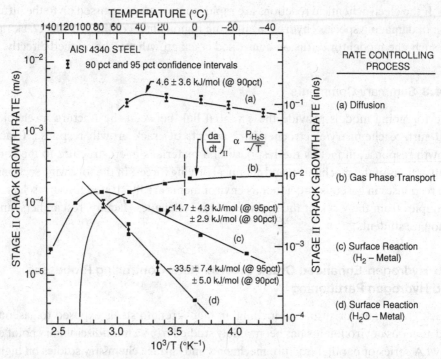

Figure 8.12. Stage II crack growth response for an AISI 4340 steel in hydrogen sulfide (a and b), hydrogen (c), and water (d) [3].

reflect the rate-controlling process, and the change in the partitioning of hydrogen between the prior austenite and martensite boundaries and the matrix, with changes in temperature. Because the embrittlement reaction, involved in the rupture of the metal-hydrogen-metal bonds, is apparently much faster, models for this final process cannot be demonstrated through correlations with experimental data.

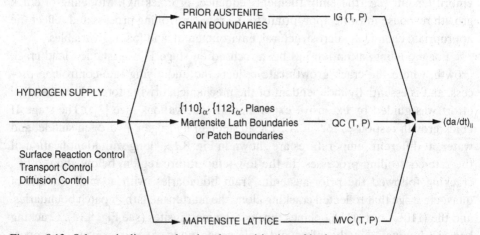

Figure 8.13. Schematic diagram showing the partitioning of hydrogen among potential paths through the microstructure [3].

Herein, results of the modeling effort are summarized, and readers are referred to [3] for specific details and references to the underlying experimental work. In essence, the model weds a microstructurally based hydrogen-partitioning function to the chemically based model for the rate of supply of hydrogen to the embrittlement zone. For simplicity, the partitioning of hydrogen is approximated in terms of its distribution between the prior austenite boundaries (*i.e.*, intergranular separation), denoted by the subscript b, and the martensite lattice (*i.e.*, microvoid coalescence), denoted by the subscript l, in the following equation.

$$\left(\frac{da}{dt}\right)_{II} \approx f_b \alpha_b \dot{Q}_b + f_l \alpha_l \dot{Q}_l = (f_b \alpha_b \kappa_b + f_l \alpha_l \kappa_l)\, \dot{Q} \qquad (8.20)$$

Equation 8.20 is a simple "rule of mixture" representation of parallel processes, whereby the overall crack growth rate is given by the sum of the fractional contribution from each of the processes; here, by intergranular cracking and microvoid coalescence, in terms of their areal fractions f_b and f_l, where $f_b + f_l = 1$. The quantities α_b, α_l, $\dot{Q}_b$, and $\dot{Q}_l$ are the proportionality constants between crack growth rate and the rate of supply of hydrogen to each of the cracking modes. The quantity $\dot{Q}$ is the total rate of hydrogen supplied to the fracture process zone, and κ_b and κ_l are the fraction of hydrogen delivered to each mode, where $\kappa_b + \kappa_l = 1$. These distribution coefficients are related to the ratio of local concentrations of hydrogen in the prior austenite grain boundaries and the matrix, and the volume fraction of these boundaries.

By using Boltzmann statistics (for dilute solutions) for the partitioning of hydrogen between the grain boundaries and the lattice, incorporating a "nonequilibrium" parameter τ to recognize that equilibrium might not be established even at steady state, the stage II crack growth rate is given by [3]:

$$\left(\frac{da}{dt}\right)_{II} = \left(\sum_i \alpha_i f_i \kappa_i\right) \dot{Q} \qquad (8.21)$$

$$= \left\{ \frac{\tau \alpha_b f_b \delta (a^3/n) N_x\ \exp(H_B/RT)}{1 + \tau \delta (a^3/n) N_x\ \exp(H_B/RT)} + \frac{\alpha_l (1 - f_b)}{1 + \tau \delta (a^3/n) N_x\ \exp(H_B/RT)} \right\} \dot{Q}$$

In Eqn. (8.21), the additional terms are: a, the lattice parameter; n, the number of atoms per unit cell; N_x, the density of trap sites in the grain boundaries; δ, the volume fraction of prior-austenite grain boundaries; H_B, the binding enthalpy of hydrogen to the grain boundary; R, the universal gas constant; and T, the absolute temperature. By explicitly incorporating the hydrogen supply rate for each of the rate-controlling processes, the corresponding crack growth rates are as follows [3]:

Transport control

$$\left(\frac{da}{dt}\right)_{II} = \left(\sum_i \alpha_i f_i \kappa_i\right) \eta_t \left(\frac{p_o}{T^{1/2}}\right) \qquad (8.22a)$$

Surface reaction control

$$\left(\frac{da}{dt}\right)_{II} = \left(\sum_i \alpha_i f_i \kappa_i\right) \eta_s p_o^m \exp\left(-\frac{E_s}{RT}\right) \tag{8.22b}$$

Diffusion control

$$\left(\frac{da}{dt}\right)_{II} = \left(\sum_i \alpha_i f_i \kappa_i\right) \eta_d p_o^{1/2} \exp\left(-\frac{E_d}{2RT}\right) \tag{8.22c}$$

Here, p_o is the external pressure; E_s and E_d are the activation energies for surface reaction and diffusion, respectively; and the η parameters relate the hydrogen supply rate to the pressure and temperature dependencies of the controlling process. Modified forms of these equations were derived by taking the parameter τ as being proportional to the hydrogen supply rate, and are also given in [3]. The efficacy of the model in predicting the temperature and pressure dependence is illustrated by the set of data for hydrogen sulfide in Fig. 8.14. For life prediction and reliability analysis, a set of key *internal* and *external* variables might be readily identified.

For comparison, the influence of temperature on stage II crack growth rates in a 18Ni (250 ksi yield strength) maraging steel, in dehumidified hydrogen at 12, 28, 57, and 133 kPa is shown in Fig. 8.15 [6, 7]. At the lower temperatures (below about 250 K), crack growth is controlled by the rate of surface reaction of hydrogen with the clean metal surfaces at the crack tip. Here, the very abrupt decreases in growth

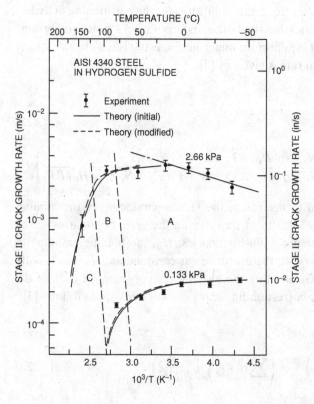

Figure 8.14. Comparison between model predictions and data for AISI 4340 steel tested in hydrogen sulfide (at 0.133 and 2.66 kPa) [3].

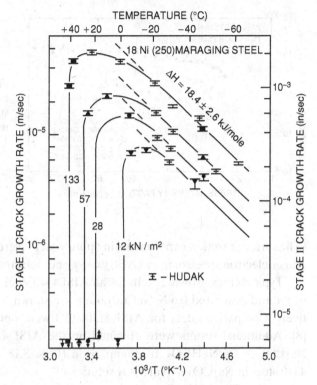

Figure 8.15. Effect of temperature on the stage II crack growth rate for 18Ni (250) maraging steel tested over a range of hydrogen pressures [6, 7].

rates with increases in temperature (*vis-à-vis*, the response of the AISI 4340 steel) could not be attributed wholly to the partitioning of hydrogen between the austenite boundaries and the martensite phases. A grain boundary phase transformation model, involving a dilute and a condensed phase, had to be invoked to explain the observed behavior. Indeed, crack growth response for the AISI 4340 steel may also involve this phase transformation in the higher temperature side of region C (unfortunately, however, the data did not extend into this region).

8.6 Electrochemical Reaction-Controlled Crack Growth (Hydrogen Embrittlement)

In the previous section, the influence of hydrogen on crack growth was clearly demonstrated. Up through the late 1970s and early 1980s, however, there was significant debate over the appropriate "mechanism" for stress corrosion cracking in aqueous environments. The essence of the debate is in the realm of the appropriate mechanism for stress corrosion cracking (SCC), or environmentally enhanced crack growth. From the corrosion perspective, SCC is the result of elcctrochemically induced metal dissolution (namely, the anodic half of the coupled reactions) at the crack tip. Hydrogen evolution is simply the other half of the coupled reaction, and is not deemed responsible for the enhancement of crack growth. A series of experiments were carried out at Lehigh University to measure crack growth kinetics, and the kinetics of bare surface reactions using an *in situ* fracture technique [8].

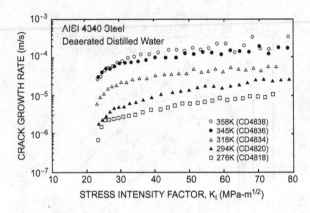

Figure 8.16. Sustained-load crack growth kinetics for AISI 4340 steel in distilled water at several temperatures [8].

(The surface reaction experiments in "pure water" were carried out separately in an Auger electron spectrometry (AES)/x-ray photoelectron spectroscopy (XPS) unit.)

Typical crack growth results for an AISI 4340 steel, in deaerated distilled (pure) water and deaerated 0.6 N NaCl solution are shown in Figs. 8.16 and 8.17, respectively. Comparable data for AISI 4130 steel were obtained and may be found in [8]. Additional results were obtained on the AISI 4130 and 4340 steels in 1 N Na_2CO_3 + 1N $NaHCO_3$, for comparison (Figs. 8.18 and 8.19), and on the AISI 4340 steel in Na_2CO_3 + $NaHCO_3$ solutions to examine the influence of anion concentration (Fig. 8.20). The apparent activation energies for crack growth and for electrochemical reaction were determined, and are shown in Table 8.1. (Note that the activation energy for reactions with pure water could not be measured electrochemically, and was estimated from surface chemical measurements [10].) Scanning electron microscope (SEM) microfractographs of AISI 4130 and 4340 steel specimens, tested in 0.6 N NaCl solution, are shown in Fig. 8.21 and show no indications of metal dissolution.

Comparison of the apparent activation energies for electrochemical reactions [8] and that of stage II crack growth (Table 8.1) show that they are equal (at the

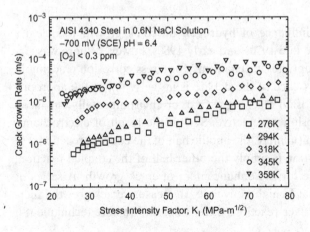

Figure 8.17. Sustained-load crack growth kinetics for AISI 4340 steel in 0.6 N NaCl solution at several temperatures [8].

Table 8.1. *Comparison of activation energies for crack growth versus electrochemical reactions*

| Environment | Crack growth | | Electrochemical/ |
	4130 steel	4340 steel	Chemical reactions
Distilled water	27 ± 11	37 ± 5	36 ± 28*
0.6 N NaCl	34 ± 7	35 ± 9	35 ± 6
1 N Na$_2$CO$_3$ + 1 N NaHCO$_3$	40 ± 13	44 ± 3	37 ± 9
Pooled	34 ± 4	38 ± 3	35 ± 3

* From 4340/water vapor reaction measurement [24].

Figure 8.18. Influence of anion on sustained-load crack growth kinetics for AISI 4130 steel at room temperature [8].

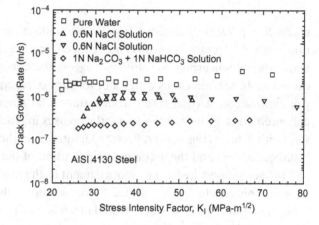

Figure 8.19. Influence of anion on sustained-load crack growth kinetics for AISI 4340 steel at room temperature [8].

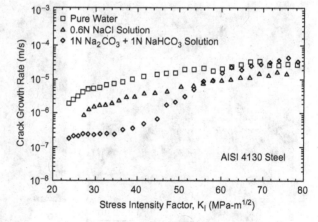

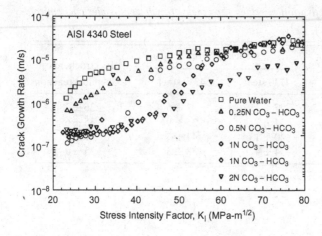

Figure 8.20. Influence of anion ($CO_3 - HCO_3$) concentration on sustained-load crack growth kinetics [8].

ninety-five percent confidence level), and confirms surface/electrochemical reaction control of crack growth. The reduced crack growth rates in the chloride and carbonate-bicarbonate solutions suggest the competition of the chloride and carbonate-bicarbonate ions with water for surface reaction sites, and support hydrogen embrittlement (that result from water-metal reaction) as the mechanism for enhanced crack growth. The observed increases in crack growth rates in AISI 4340 steel with K level (Fig. 8.20) reflects the limitation in the transport of the carbonate-bicarbonate ions and the accompanying dilution of the electrolyte at the crack tip, and further support hydrogen embrittlement as the mechanism for enhancing crack growth. This conclusion is affirmed by the scanning electron microfractographs of AISI 4130 and AISI 4340 steels, tested in 0.6 N NaCl solution (Fig. 8.21), that show no evidence of electrochemical dissolution of the crack surfaces.

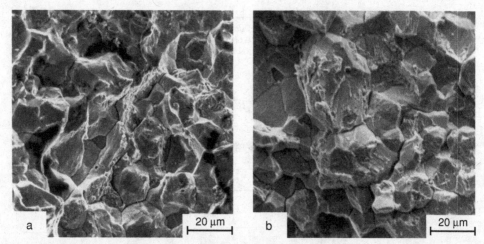

Figure 8.21. Fracture surface Morphology for sustained-load crack growth in 0.6 N NaCl solution at K = 33 MPa-m$^{1/2}$ and 294 K: (a) AISI 4130 steel, and (b) AISI 4340 steel [8].

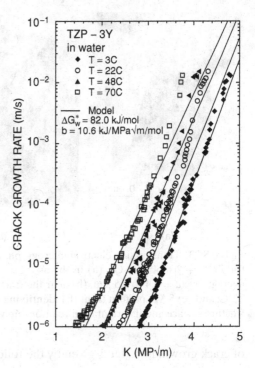

Figure 8.22. Crack growth data for TZP-3Y zirconia (ZrO_2 + 3 mol% Y_2O_3) in water. Solid lines represent model predictions [4].

8.7 Phase Transformation and Crack Growth in Yttria-Stabilized Zirconia

To better understand environmentally enhanced crack growth in yttria-stabilized zirconia (ZrO_2 + 3 mol% Y_2O_3), a series of experiments was conducted to determine the kinetics of crack growth and associated changes in microstructure [9]. Crack growth tests under a statically applied load were conducted in water, dry nitrogen, and toluene from 276 to 343 K. Transformation induced by moisture (water) and stress was determined by postfracture examination of the region near the fracture surfaces by x-ray diffraction analyses and transmission electron microscopy. These microstructural examinations were supplemented by studies of stress-free specimens that had been exposed to water at the higher temperatures. Data on the kinetics of crack growth in water (*i.e.*, the individual data points) are shown in Fig. 8.22, and evidence for phase transformation during crack growth is shown in Fig. 8.23. The results, combined with literature data on moisture-induced phase transformation, suggested that crack growth enhancement by water is controlled by the rate of this tetragonal-to-monoclinic phase transformation and reflects the environmental cracking susceptibility of the resulting monoclinic phase.

By assuming that the rate of crack growth is controlled by the rate of tetragonal-to-monoclinic phase transformation, a kinetic model was proposed as an analogue to that for martensitic transformation. Only the final form of the model is given here; specific details of its formulation may be found in [9]. In this model, the rate

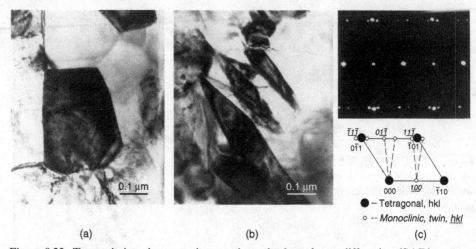

Figure 8.23. Transmission electron micrographs and selected area diffraction (SAD) pattern for ZrO_2 + 3 mol% Y_2O_3: (a) in the as-received condiition showing equiaxed grains with average size $d = 0.4$ to 0.5 μm; (b) near the fracture surface of a specimen tested in water at $22°C$; and (c) SAD pattern from (b) identifying the new twinned martensite phase near the fracture surface and its orientation relationship with the t-matrix [4].

of crack growth in water is given by the following equation:

$$\frac{da}{dt} = \dot{A}_{ow} \exp\left(-\frac{\Delta G_w^* - \Delta G_K^*}{RT}\right)$$

$$= \dot{A}_{ow} \exp\left(-\frac{\Delta G_w^* - (\alpha K_I)V_w^*}{RT}\right)$$

(8.23)

In Eqn. (8.23), $\dot{A}_{ow}$ is currently an experimentally determined rate constant, which would depend on the microstructure and its interaction with water; ΔG_w^* is the effective activation energy barrier for the tetragonal to monoclinic phase transformation in water; and ΔG_K^* is the reduction in the activation energy barrier for phase transformation by the crack-tip stresses. The term ΔG_K^* is given in terms of the stress-enhanced strain energy density for transformation, αK_I, and an activation volume, V_w^*, where K_I is the crack-tip stress intensity factor for mode I loading [9]. Comparison of the model with the experimental data, after establishing the rate constant from the data at one test temperature, is shown by the straight lines in Fig. 8.22. Departures at the higher K_I levels correspond to the onset of crack growth instability, and are not represented by the model.

The various material-related terms in Eqn. (8.23) are considered to be *internal* variables. Because their expected dependence on composition and microstructure (*i.e.*, on the concentration of yttria and volume fraction of the tetragonal phase), they are to be viewed as random variables. The rate constant $\dot{A}_{ow}$ is expected to depend on the mechanism for the enhanced tetragonal-to-monoclinic phase transformation by water, perhaps the replacement of Zr-O-Zr bonds by OH bonding, and needs to be better understood and quantified.

8.8 Oxygen-Enhanced Crack Growth in Nickel-Based Superalloys

This section reflects the more recent venture by the author and his colleagues into the realm of environmentally enhanced crack growth at high temperatures. The embrittling agent here is oxygen, *vis-à-vis*, hydrogen, for alloys at the lower temperatures, in gaseous hydrogen, and in aqueous environments. This investigation required the use of sophisticated surface analysis tools and well controlled experiments. The following subsections summarize the procedures, key results, and consequent new understanding of the role of oxygen in crack growth enhancement that has been derived.

Nickel-based superalloys, such as IN 100 and Inconel 718, are used extensively in high-temperature applications in oxidizing environments; for example, as disks in turbine engines. The influence of oxygen and moisture on crack growth in these alloys at high temperatures has been recognized for a long time. The presence of oxygen can increase the rate of crack growth under sustained loading by up to 4.5 orders of magnitude over that in inert environments. Considerable efforts have been made to understand the mechanisms for this enhancement, for example, [11–20]. Floreen and Raj [11] have categorized the various mechanisms into two groups. The first group involves environmentally enhanced formation and growth of cavities, or microcracks, at grain boundaries ahead of the crack tip. The second type is associated with preferential formation of a grain boundary oxide layer at the crack tip. Specifically, the mechanisms include: (a) the oxidation of metallic carbides or carbon at the grain boundaries to form CO and CO_2 gases at high internal pressures to enhance cavity growth along grain boundaries, (b) the nucleation and growth of Ni and Fe oxides directly behind the crack tip during propagation to form an oxide "wedge," and (c) the formation of Ni oxides directly behind the crack tip at high oxygen pressures, while Cr oxidizes at low pressures to inhibit alloy failure [18–20].

More recent studies [21–32] on an Inconel 718 alloy (under sustained loading) and a Ni-18Cr-18Fe ternary alloy (in fatigue) suggested that niobium (Nb) can play a significant role in oxygen-enhanced crack growth (OECG), and raised concerns regarding the viability of these proposed mechanisms for crack growth enhancement by oxygen. The results showed, for example, that the crack growth rates under sustained load in oxygen at 973 K were more than 10^4 times higher than those in high-purity argon for Inconel 718 [21, 22]. The enhancement in crack growth rate was attributed to the formation and rupture of a nonprotective and brittle Nb_2O_5-type oxide film at the grain boundaries through the oxidation and decomposition of Nb-rich carbides and, perhaps, oxidation of γ'' (Ni_3Nb) precipitates at the grain boundaries [22, 30, 31]. Crack growth rates in the ternary alloy, on the other hand, were found to be essentially unaffected by oxygen [32]. A comparison of these results with those on a range of nickel-based superalloys in the literature showed a strong dependence of the environmental sensitivity factor (*i.e.*, the ratio of crack growth rates in the deleterious and inert environments) on Nb concentration [32]. The sensitivity factor increased by more than 10^4 times with increases in Nb from zero to five weight percent; *albeit* the sensitivity varied among the alloys with a given Nb concentration.[2] These findings suggest that the role of Nb on enhancing crack growth in oxygen, heretofore not recognized, needs to be carefully examined. The "insensitivity" of the Ni-18Cr-18Fe ternary alloy, with copious amounts of $M_{23}C_6$ carbides at the grain boundaries, calls into question the viability of both groups of previously proposed mechanisms.

Here, the results from a series of coordinated crack growth, microstructural, and surface chemistry studies to elucidate the role of niobium and other alloying elements (such as Al and Ti) on crack growth in oxygen at high temperatures are summarized [33–38]. These studies complement the earlier work on Inconel 718 [21–31]. Three γ'-strengthened powder metallurgy (P/M) alloys, with nominal composition similar to alloy IN-100, but with 0, 2.5, and 5 weight percent Nb (designated as alloys 1, 2 and 3, respectively), were investigated. These alloys were designed to suppress the formation of γ'' (and δ) precipitates, such that the impact of Nb-rich carbides (*vis-à-vis* Ni_3Nb) could be separately identified.

8.8.1 Crack Growth

Crack growth data were obtained, for the circumferential-radial (CR) orientation, under constant load in high-purity oxygen at 873, 923, and 973 K [37, 38]. Because of the very slow rates of crack growth (less than 10^{-5} meters per hour), testing in high-purity argon was limited principally to the intermediate levels of the mechanical driving force K (*i.e.*, about 60 MPa-m$^{1/2}$) at 973 K. The crack growth rate versus K results for alloys 1 and 3 are shown in Figs. 8.24 and 8.25, respectively, as a function of temperature [37, 38]. The crack growth rates and response for alloy 2 (not shown) are essentially identical to those of alloy 3 (Fig. 8.25). The crack growth rates and responses in these Nb-containing alloys parallel those of Inconel 718, and their

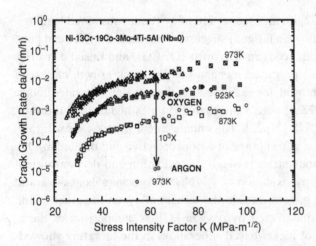

Figure 8.24. The kinetics of crack growth for alloy 1 in high-purity oxygen and argon at 873, 923, and 973 K [37, 38].

---

[2] The trend line and indicated variability in [32] may have underestimated the environmental sensitivity, because the reference environment in the earlier studies may not be sufficiently low in oxygen and moisture.

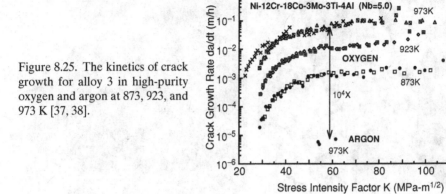

Figure 8.25. The kinetics of crack growth for alloy 3 in high-purity oxygen and argon at 873, 923, and 973 K [37, 38].

growth rates at 973 K were about 10^4 times faster than those in argon ($p_{O2} < 10^{-17}$ Pa) [21, 22, 37]. Results on the Nb-free alloy 1, however, were surprising (*cf.* Figs. 8.24 and 8.25) in that the growth rates in oxygen were nearly 10^3 times faster than those in argon at 973 K (see Fig. 8.24); *i.e.*, with an environmental sensitivity factor that is well above one time to ten times anticipated from the literature data [22].

Crack growth rates in oxygen strongly depend on temperature and appear to be controlled by a thermally activated process that depends on the K level. The apparent activation energy for crack growth for each of the alloys was estimated on the basis of steady-state crack growth rates at K levels from 35 to 60 MPa-m$^{1/2}$. For the Nb-free alloy 1, the apparent activation energy for crack growth was essentially constant at about 250 kJ/mol. For the Nb-containing alloys (alloys 2 and 3), on the other hand, it decreased from about 320 to 260 kJ/mol with increasing K from 35 to 60 MPa-m$^{1/2}$ [37, 38]. The consistency between the apparent activation energy for crack growth in the alloys, and with those reported for other nickel-based superalloys [22, 37, 38], suggests that a common process controlled the rate of crack growth. A plausible rate-controlling process for crack growth is that of stress-enhanced diffusion of oxygen into the crack-tip process zone. The K dependence, observed in the Nb-containing alloy versus the Nb-free alloy, is likely a reflection of the influence grain size (45 versus 10 μm), and requires further study.

Representative microfractographs of the Nb-free alloy 1 and Nb-containing alloy 3 tested in high-purity oxygen, under sustained load, at 973 K are shown in Fig. 8.26 [37, 38]. Cracking in alloy 1 was essentially intergranular and also followed along interfaces between the large (micrometers in size) primary γ' precipitates and the matrix, some of which are indicated by the arrows in Fig. 8.26a. For alloy 3, Fig. 8.26b, cracking was predominately intergranular. Many small particles (appearing in white) are seen on the grain boundary surfaces, and were found through energy dispersive spectroscopy (EDS) analyses to be rich in Nb. These particles are consistent with the Nb-rich carbides found previously on the grain boundaries of Inconel 718 [31].

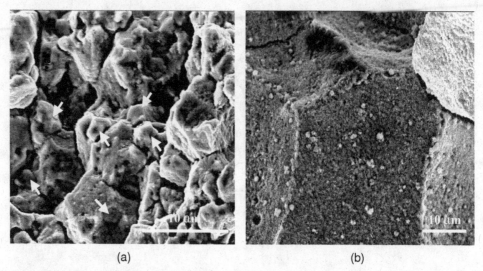

(a)   (b)

Figure 8.26. Microfractographs of (a) Nb-free alloy 1, with arrows pointing to primary γ', and (b) Nb-containing alloy 3, tested in oxygen at 973 K [37, 38].

8.8.2 High-Temperature Oxidation

The reactivity of the alloys to oxygen was determined on polished and ion-sputtered surfaces of alloys 1 and 3, as well as Inconel 718 by x-ray photoelectron spectroscopy (XPS) in vacuum at 10^{-7} Pa, and following exposure to 5×10^{-4} Pa of oxygen at 873, 923, and 973 K for 2,700 s [24–26]. Specimens of Nb, Ni_3Nb (with γ'' crystal structure), and a specially grown film of NbC were given the same exposure to oxygen at 873, 923, and 973 K and analyzed by XPS to provide direct evidence for the reactions of Nb compounds with oxygen, [33].

For alloys 1 and 3 in ultrahigh vacuum (UHV) and 5×10^{-4} Pa O_2 at 973 K, the XPS peak areas for each elemental region were normalized by using their corresponding sensitivity factors. The normalized, relative-peak-area percentages for the sputtered and reacted surfaces of these alloys are shown in Table 8.2 in

Table 8.2. *Normalized, relative-XPS-peak-area percentage for sputtered surfaces of alloys 1 and 3, and of Inconel 718 after heating in UHV for sixty minutes and 5×10^{-4} Pa O_2 for forty-five minutes at 973 K [33, 34].*

	Ni	Cr	Co	Fe	Ti	Mo	Al	Nb
Alloy 1 (sputtered)	51.4	10.5	21.0	–	2.8	4.0	10.3	–
UHV (973 K)	31.7	19.8	17.3	–	5.7*	3.9	21.6*	–
Oxygen (973 K)	8.0	44.8*	3.8	–	10.6*	2.0	30.9*	–
Alloy 3 (sputtered)	52.8	10.3	20.2	–	1.5	4.2	7.1	4.0
UHV (973 K)	37.4	16.2	16.6	–	3.3*	4.8	15.5*	6.1*
Oxygen (973 K)	14.5	40.6*	4.4	–	5.8*	2.1	20.9*	11.8*
718 (sputtered)	59.7	15.9	–	12.9	0.9	3.8	3.1	3.6
UHV (973 K)	34.5	35.1*	–	9.3	1.5*	4.3	5.9*	9.4*
Oxygen (973 K)	2.3	64.3*	–	20.9*	0.9*	0.5	2.1*	9.0*

* oxidation of the element.

comparison with Inconel 718 [33, 34]. For the sputtered surfaces of alloys 1 and 3 and of Inconel 718 heated in UHV at 973 K, Al, Cr, Ti, and Nb (in alloy 3 and Inconel 718) enriched the surface, whereas Ni decreased. The surface Al and Ti on all three alloys oxidized significantly to Al_2O_3 and TiO. Niobium on alloy 3 oxidized to a lesser extent to NbO_x ($1 < x < 2$), whereas it oxidized significantly to Nb_2O_5 on Inconel 718 [33–36]. Oxidation of these sputtered surfaces during heating in UHV is most likely caused by the evolution and migration of dissolved atomic oxygen in the sample to the polished surface, and by the exposure to O_2 and H_2O that had out-gassed from the heater and other heated parts of the system. After exposure to oxygen (5×10^{-4} Pa) at 973 K, Cr, Ti, Al, and Nb (on alloy 3 and Inconel 718) enriched the surface and oxidized to form thick films of Al_2O_3, Cr_2O_3, TiO_2, and Nb_2O_5, respectively. No substrate signal was observed, signifying that the oxide film thickness was greater than the escape depth of the photoelectrons of approximately ten nanometers. Nickel, on the other hand, remained unoxidized under either of the conditions. These findings are consistent with the preferential oxidation of Al and Ti in Ni_3Al and Ni_3Ti to form Al_2O_3 and TiO_2, which has been confirmed by the oxidation of these intermetallic compounds over the range of conditions in a separate study [26]. Chromium did not oxidize during heating in UHV. It was enriched and oxidized to Cr_2O_3, however, after heating to 973 K in 5×10^{-4} Pa O_2, which suggests the oxidation of $Cr_{23}C_6$, or Cr in solid solution, or both.

The relative reactivity and extent of reactions of Nb, Ni_3Nb, and NbC with O_2 were determined by XPS [33]. Specimens following exposure to 5×10^{-4} Pa of oxygen at 873, 923, and 973 K for 2,700 seconds were analyzed. Niobium, Ni_3Nb, and NbC were found to oxidize readily. Representative spectra for Ni_3Nb and the NbC film at 973 K are shown in Fig. 8.27 along with the component spectra for Ni_3Nb and

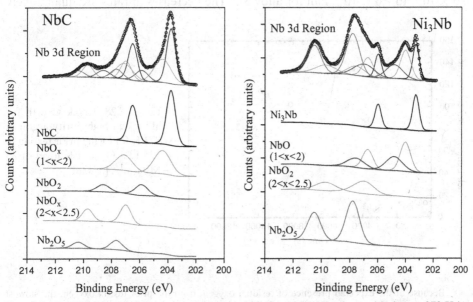

Figure 8.27. XPS spectra (Nb region: $3d_{3/2,5/2}$) of NbC and Ni_3Nb after oxidation at 973 K in 5×10^{-4} Pa of O_2 for forty-five minutes [33].

NbC and the various oxides that were formed. The results confirm the fact that both Ni_3Nb and NbC react with oxygen at these temperatures to form niobium oxides of ranging stoichiometry, with a greater propensity for Ni_3Nb to oxidize to Nb_2O_5.

Additional experiments on Ni_3Al and Ni_3Ti confirmed that these precipitates also oxidized readily. The oxidation of these precipitates ahead of the crack tip contributed to the enhancement of crack growth in the γ'-strengthened alloys [36, 38]. The increase in growth rate in the γ'-strengthened alloys appears to depend on the volume fractions of these precipitates [38].

8.8.3 Interrupted Crack Growth

To ascertain the existence of an OAR ahead of the tip of a growing crack and to determine which elements might be oxidized there, mechanically and chemically based, interrupted crack growth experiments were carried out. The findings from these experiments are summarized below.

8.8.3.1 Mechanically Based (Crack Growth) Experiments

To ensure experimental control, the interrupted experiments were carried out on alloys 1 and 3 at 873 K in 135 kPa O_2. Sustained-load crack growth in oxygen was interrupted at K $\approx$ 33 MPa-m$^{1/2}$. The specimen was partially unloaded and the test chamber was evacuated and back-filled with ultrahigh purity argon, and was then reloaded to the same test load. Crack growth responses for these alloys are shown in Figs 8.28 and 8.29 [38]. The results attest to the presence of an embrittled zone (or oxygen-affected region (OAR)) of about 80 μm for each alloy. The crack growth rates over this zone decreased from their preinterruption rates in oxygen to that for argon (*i.e.*, from about 5.8×10^{-5} to $<1.5 \times 10^{-5}$ m/h for alloy 1, and from about 6.6×10^{-5} to $<6 \times 10^{-6}$ m/h for alloy 3).[3] The decreases in rates are qualitatively

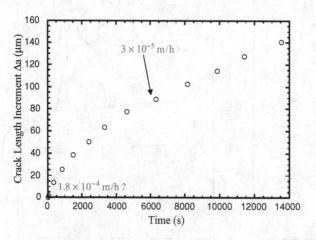

Figure 8.28. Crack growth for alloy 1 in high-purity argon at 873 K following prior testing in 135 kPa O_2 at a K level of 33 MPa-m$^{1/2}$ [37, 38].

---

[3] Because of the expected presence of residual oxygen, from the prior test in oxygen, the slowest crack growth rates here would tend to be higher than the corresponding rates in high-purity argon.

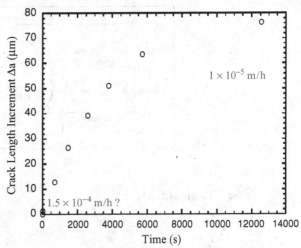

Figure 8.29. Crack growth for alloy 3 in high-purity argon at 873 K following prior testing in 135 kPa O_2 at a K level of 33 MPa-m$^{1/2}$ [37, 38].

consistent with the expected variation in oxygen concentration profile with diffusion, and are attributed to the penetration of oxygen ahead of the crack tip ($\sim$80 μm) and internal oxidation of grain boundary surfaces during the preinterruption crack growth in oxygen. The presence of an OAR suggests that there should be chemical manifestations of its presence; this finding is examined and compared with results from the chemically based experiments given in the following subsection.

8.8.3.2 Chemically Based Experiments (Surface Chemical Analyses)

To establish the presence of an OAR ahead of the crack tip chemically and to determine which elements might react with oxygen there, the near-tip fracture surfaces from the interrupted crack growth test specimens of alloys 1 and 3 were analyzed by XPS [35, 36]. The results are compared with those from Inconel 718 [33]. The regions of the fracture surfaces used for analysis are illustrated schematically in Fig. 8.30a, and representative SEM microfractographs of this region for alloy 1 and alloy 3 are shown in Fig. 8.30b and 8.30c, respectively. The "*in situ* fracture surface" is the area (ahead of the crack tip) that was exposed by *in situ* fracture in UHV and includes the OAR. It is represented by the transgranular fracture region on the left side of Fig. 8.30b and 8.30c, and a small region of the adjacent intergranular fracture (the OAR). The "fracture surface" is the area (behind the crack tip) that is exposed during crack growth in 135 kPa O_2 at 973 K; *i.e.*, the bulk of the intergranular fracture region seen on the right side of Fig. 8.30b and 8.30c. The OAR is defined chemically by the area that is oxidized ahead of the tip of the growing crack, and the limit of the OAR is the furthest point of this oxidation. The XPS analyses were performed at 50-μm intervals along the crack growth direction, using an automated sample manipulator and an estimated analysis window of 75 by 400 μm.

The normalized, relative-peak-area percentages as a function of distance for the fracture surfaces of alloy 1 are shown in Fig. 8.31a, and for alloy 3 in Fig. 8.32a. The 0-μm marker corresponds to the starting position for the analyses, and the 700-μm region encompassed the analyzed areas ahead and behind the crack tip. The

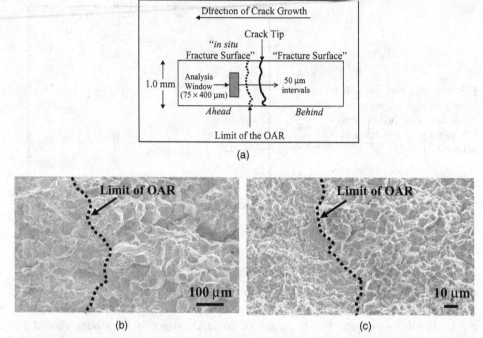

Figure 8.30. (a) Schematic representation of the crack-tip region of the P/M alloys analyzed by XPS. Representative SEM microfractographs of surfaces produced during crack growth in oxygen and subsequently by *in situ* fracture for (b) alloy 3 and (c) alloy 1 [35, 36].

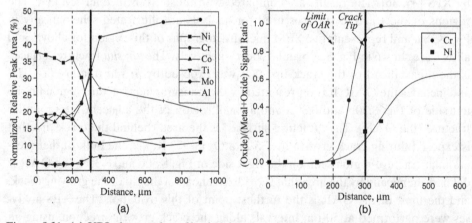

Figure 8.31. (a) XPS elemental profile showing the normalized, relative peak area percentage for fracture surfaces of alloy 1 versus distance, and (b) the normalized oxidation profile showing the oxide formation for Cr and Ni on fracture surfaces of alloy 1 versus distance [35, 36].

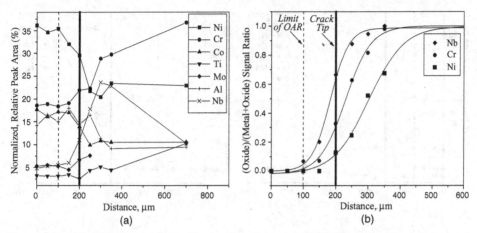

Figure 8.32. (a) XPS elemental profile showing the normalized, relative peak area percentage for fracture surfaces of alloy 3 versus distance, and (b) the normalized oxidation profile showing the oxide formation for Nb, Cr, and Ni on fracture surfaces of alloy 3 versus distance [35, 36].

elemental profiles show a substantial increase in Cr and Al beginning at the 200-μm marker for alloy 1, and in Cr and Nb, beginning at the 100-μm marker for alloy 3, while Ni and Co decreased. To determine the location and extent of the OAR ahead of the crack tip, the amount of Nb, Cr, and Ni oxidation was found by (a) curve fitting the XPS spectra, (b) normalizing the oxide signal for each element, and (c) fitting the data with a Gaussian curve. The fitted, normalized oxidation curves are shown in Figs. 8.31b and 8.32b for alloys 1 and 3, respectively. Because it is well known that Ni in superalloys begins to oxidize in oxygen at about 1.0 Pa and 973 K [8, 9], it was assumed that Ni oxidized at least to the crack-tip during the sustained-load crack growth experiments. Based on this assumption, the crack-tip position was estimated to correspond to the point where Ni first oxidizes. Similarly, the limit of the OAR was estimated to be, at least, at the last position ahead of the crack tip where Cr oxidized for alloy 1, and where Nb oxidizes for alloy 3. The estimated size of the OAR for both alloys is about 100 μm (see Figs. 8.31b and 8.32b). With the use of the standardized procedure, the error in estimating the size of OAR would principally reflect small differences in crack front contour within the analysis window, and is estimated to be about ±10 μm.

Aluminum and Ti were not included in the oxidation profiles in Figs. 8.31b and 8.32b because their XPS signal-to-noise ratios were too low to accurately determine the extent of their oxidization. Because Al and Ti oxidized to a greater extent during heating in UHV and have greater thermodynamic potential to oxidize in lower levels of oxygen than Nb and Cr, they are expected to be oxidized further ahead of the crack tip than Nb and Cr during crack growth [26]. The actual OAR for these alloys, therefore, is expected to be larger than the estimated 100 μm. It should be noted that Ni and Fe (not shown) were not oxidized fully within a region approximately 200 μm behind the crack tip.

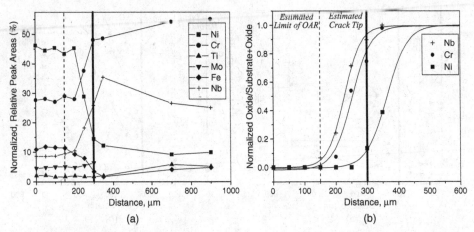

(a)   (b)

Figure 8.33. (a) XPS elemental profile showing the normalized, relative peak area percentage for fracture surfaces of Inconel 718 versus distance, and (b) the normalized oxidation profile showing the oxide formation for Nb, Cr, and Ni on fracture surfaces of Inconel 718 versus distance [35, 36].

For comparison, the relative-peak-area percentages and the normalized oxide signal of Nb, Cr, and Ni for Inconel 718 are shown in Figs. 8.32a and 8.32b, respectively [33]. The normalized, relative-peak-area percentages show a substantial increase in Cr and Nb beginning at the 150-μm marker, while Ni and Fe decrease (Fig. 8.33a). The normalized oxidation profiles for Nb, Cr, and Ni are shown in Fig. 8.33b, and indicate a larger estimated OAR of 150 μm relative to alloys 1 and 3. The ability of Nb to be oxidized further ahead of the crack tip (*i.e.*, a larger OAR) is attributed to the low concentrations of Al and Ti in this alloy and the absence of competition of these elements (*vis-à-vis* Nb) for oxygen. As such, the estimated size of OAR at $K = 60$ MPa-m$^{1/2}$ is judged to be about 150 μm (versus 100 μm) for all three alloys. Again, Ni and Fe oxidation behind the crack tip was limited.

The XPS data from the fracture surfaces of these alloys demonstrate that oxygen-penetrated ahead of the crack tip during crack growth and then oxidized Nb to Nb_2O_5 and Cr to Cr_2O_3, while Ni underwent minimal or no oxidation. These results are consistent with the high-temperature oxidation of alloy 3 and Inconel 718 in UHV (*i.e.*, Nb oxidized to a greater extent than Cr, and Ni decreased and remained unoxidized). Comparisons of the data for alloy 1 with alloy 3, and for alloy 3 with Inconel 718, suggest that Al and Ti preferentially oxidized (*vis-à-vis*, Cr and Nb) ahead of the crack tip. This suggestion is consistent with the high-temperature oxidation data on these alloys as well (Table 8.2). The results suggest that the OAR of 100 μm for alloys 1 and 3 had been underestimated because of the limitations of the XPS technique in quantifying the oxidation of Al and Ti ahead of the crack tip. The larger size of 150 μm for 973 K at $K = 60$ MPa-m$^{1/2}$ compared with the estimated OAR of 80 μm for 873 K at $K = 33$ MPa-m$^{1/2}$ from the mechanically based experiments is consistent with the expected dependence of OAR on temperature and K level.

8.8.4 Mechanism for Oxygen-Enhanced Crack Growth in the P/M Alloys

These results show that Al, Ti, Nb, and Cr (*vis-à-vis*, Ni and Fe) are oxidized ahead of the tip of a growing crack. Kinetically, Al and Ti are the most reactive, followed by Nb and then Cr. They indicate that the oxidation of Al, Ti, and Nb, and possibly Cr, are the probable cause for OECG in the nickel-based superalloys, and suggest that oxygen diffusion along grain boundaries and interfaces as the rate-controlling process. These findings are consistent with previous studies that showed: (a) Al, Ti, Cr, and Nb are internally oxidized in the alloys heated in oxygen at high temperatures [40, 41]; (b) the metal oxides tend to be more brittle than the elements themselves [32, 33]; and (c) the rate of oxygen diffusion along grain boundaries is orders of magnitude higher than through the matrix [34–36]. The oxidation of these elements ahead of the growing crack differs from the previously suggested role for Ni and Fe for oxygen-enhanced crack growth in the Ni-based superalloys [27–29].

For alloy 1, OECG resulted from the formation of brittle oxides of Al and Ti along grain boundaries and γ'-matrix interfaces ahead of the crack tip, and enhanced cracking along these embrittled boundaries and interfaces. Chemically, this mechanism is consistent with the findings that Al and Ti are readily oxidized in alloy 1, and with the preferential oxidation of these elements in Ni_3Al and Ni_3Ti [36]. Microstructurally and fractographically, it is consistent with the presence of copious amounts of secondary γ' precipitates adjacent to the grain boundaries, and cracking along the incoherent interfaces between the primary γ' precipitates and the matrix [37–39]. Because of incoherency of the primary γ'-γ interfaces, Wei [38] suggested that interfacial oxidation of the primary γ' precipitates (*albeit* at a much smaller volume fraction than the secondary precipitates) produces more severe embrittlement and contributed disproportionately to OECG rate [38].

For alloy 3, and by inference alloy 2, OECG is attributed in part to the oxidation of Al and Ti associated with the strengthening, secondary γ' precipitates adjacent to the grain boundaries. (Primary γ' precipitates were essentially absent in these alloys.) The higher crack growth rates in oxygen in these alloys relative to alloy 1 (about ten times higher at 973 K) indicate that Nb also played a significant role. Because the formation of γ' (Ni_3Nb) was suppressed, crack growth enhancement by Nb can only be attributed to the internal oxidation of Nb-rich carbides along the grain boundaries. Because of the small number of these carbides on the grain boundaries (versus γ' precipitates), a high degree of mobility of Nb over the boundary surfaces would be required. Data on the oxidation of NbC lend support to this mechanism which was originally proposed by Gao and colleagues [22, 23]. These investigators showed that rapidly heating a single crystal of Inconel 718 alloy with an adsorbed layer of oxygen to 973 K, in a thermal desorption experiment, resulted in the evolution of CO and a severalfold increase in the surface concentration of Nb. They attributed their findings to the decomposition of Nb-rich carbides at the surface of Inconel 718, and the subsequent migration of Nb over the crystal surface. They suggested that this process and subsequent oxidation of Nb along the grain boundaries is a principal mechanism for OECG in Inconel 718.

The role of the γ'' (Ni_3Nb) precipitates was not fully addressed in the previous study [21–32]. The surface reaction data herein, however, show that Nb in Ni_3Nb is preferentially oxidized at high temperatures to form niobium oxides of varying stoichiometry. As such, it is expected that the oxidation of γ'' precipitates that lie along the grain boundaries of the γ''-strengthened alloys, such as Inconel 718, would be a significant contributor to OECG. These results taken *in toto* showed that both Ni_3Nb and Nb-rich carbides contributed to OECG in the γ''-strengthened alloys, and that the oxidation of Nb-rich carbides *per se* is a significant contributor.

The role of Cr is less certain at this time and requires further study. Although its oxidation ahead of the crack tip is clearly evident, the fact that chromium forms a strong and coherent oxide, which long served as the basis for its use in stainless steels, argues against its role in OECG. The essential absence of crack growth sensitivity of the Ni18Cr18Fe ternary alloy to oxygen also questions its efficacy as an embrittler.

The size of the OAR (*i.e.*, the extent of the region of oxide coverage ahead of the crack tip) is a measure of the distance of diffusive oxygen penetration ahead of a growing crack. The crack growth and surface reaction data suggest that stress-enhanced diffusion of oxygen along the grain boundaries and γ'-matrix interfaces is the rate-controlling process for OECG. This suggestion, however, needs to be confirmed by direct measurements of diffusion.

8.8.5 Importance for Material Damage Prognosis and Life Cycle Engineering

The importance of these findings on material damage prognosis and life cycle engineering may be illustrated through a consideration of the influence of the volume, or area fraction of γ' precipitates on crack growth in the nickel-based alloys. Under sustained loads, the rate of crack growth (da/dt), at a given K level, may be given by the superposition of a creep-controlled component, $(da/dt)_{cr}$, and an environmentally affected component, $(da/dt)_{en}$, as follows:

$$\left(\frac{da}{dt}\right) = \left(\frac{da}{dt}\right)_{cr} \phi_{cr} + \left(\frac{da}{dt}\right)_{en} \phi_{en} \qquad (8.24)$$

The terms ϕ_{cr} and ϕ_{en} are the areal fractions of creep-controlled and environmentally affected crack growth, respectively. Assuming that, at K levels higher than the crack growth threshold (K_{th}), the creep crack growth rate may be neglected (compared with the corresponding growth rates in oxygen; Figs. 8.24 and 8.25), the growth rate is then essentially that of the environmentally affected component, and is given by Eqn. (8.25):

$$\left(\frac{da}{dt}\right) \approx \left(\frac{da}{dt}\right)_{s\gamma'} f_{V\gamma'}; \quad \text{for } f_{V\gamma'} \leq f_{V\gamma'}^*$$

$$\left(\frac{da}{dt}\right) \approx \left(\frac{da}{dt}\right)_{s\gamma'} f_{V\gamma'}^* + \left(\frac{da}{dt}\right)_{p\gamma'} \left(f_{V\gamma'} - f_{V\gamma'}^*\right); \quad \text{for } f_{V\gamma'} \leq f_{V\gamma'}^*$$

$$(8.25)$$

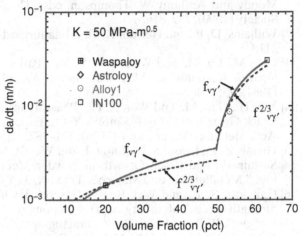

Figure 8.34. Estimated dependence of oxygen-enhanced crack growth rates on γ' volume fraction in γ'-strengthened nickel-base alloys [37].

Here, $(da/dt)_{sy'}$ and $(da/dt)_{py'}$ are the rates of crack growth associated with the secondary and primary precipitates, and the areal fraction of precipitates is assumed to be equal to the volume fraction, and $f^*_{V\gamma'}$ is the volume fraction of γ' at the onset of coarsening (or the transformation from secondary to primary precipitates). Alternatively, the area fraction may be taken to be equal to the 2/3-power of the volume fraction. The possible dependence of sustained-load crack growth rates on γ' volume fraction, in oxygen at $K = 50$ MPa-m$^{1/2}$, is shown in Fig. 8.34 [37, 38]. The volume fraction of γ' precipitates in Waspaloy and the associated trend line for crack growth rates are estimates, whereas those at the higher volume fractions are supported by data [37, 38].

8.9 Summary Comments

In this chapter, the disciplines and processes that need to be brought to bear to solve *real* problems that involve materials in *realistic* environments are highlighted through selected examples. The development and use of tools for design and management of engineered systems must incorporate mechanistically based understanding and modeling of material response in terms of loading (stress analyses), environmental, and microstructural variables. The foregoing discussions have dealt with cracking (crack growth) from a linear elastic fracture mechanics-based perspective. It is recognized that many machines and structures are subject to "cyclic" or "fatigue" loading. The impact of these loads on damage evolution and service life prediction is discussed in Chapter 9. Note also that many materials can undergo corrosion damage (both in and out of service) and hasten crack nucleation and significantly shorten service life. Such interactions are introduced in brief in Chapter 10.

REFERENCES

[1] Wei, R. P., and Gao, M., "Hydrogen Embrittlement and Environmentally Assisted Crack Growth," in *Hydrogen Effects on Material Behavior*, Neville R.

Moody and Anthony W. Thompson, eds., The Minerals, Metals & Materials Society (1990), 789–816.

[2] Williams, D. P., and Nelson, H. G., Metallurgical Transactions, 3 (1972), 2107–2113.

[3] Gao, M., Lu, M., and Wei, R. P., "Crack Paths and Hydrogen-Assisted Crack Growth Response in AISI 4340 Steel," Metallurgical Transactions A, 15A (1984), 735–746.

[4] Yin, H., Gao, M., and Wei, R. P., "Phase Transformation and Sustained-Load Crack Growth in ZrO_2 + 3 mol% Y_2O_3: Experiments and Kinetic Modeling," Acta Metall. et Mater., 43, 1 (1995), 371–382.

[5] Huang, Z. F., Iwashita, C., Chou, I, and Wei, R. P., "Environmentally Assisted, Sustained-Load Crack Growth in Powder Metallurgy Nickel-Based Superalloys," Metallurgical and Materials Trans A, 33A (2002), 1681–000.

[6] Gangloff, R. P., and Wei, R. P., "Gaseous Hydrogen Embrittlement of High Strength Steels," Metallurgical Transactions A, 8A (1977), 1043–1053.

[7] Gangloff, R. P., and Wei, R. P., "Fractographic Analysis of Gaseous Hydrogen Induced Cracking in 18Ni Maraging Steel," Fractography in Failure Analysis, ASTM STP 645 (1978), 87–106.

[8] Chu, H. C., and Wei, R. P., Corrosion, 46, 6 (1990), 468–476; Chu, H. C., Dissertation, Lehigh University (1987).

[9] Alavi, A., Miller, C. D., and Wei, R. P. Corrosion, 43, 4 (1987), 207.

[10] Simmons, G. W., Pao, P. S., and Wei, R. P., Met. Trans. A, 9A (1978), 1147.

[11] Floreen, S., and Raj, R. in Flow and Fracture at Elevated Temperatures, R. Raj, ed., Am. Soc. Metals, Metals Park, OH (1984), 383.

[12] Woodford, D. A., and Bricknell, R. H., Acta Met. 30 (1982), 257.

[13] Woodford, D. A., and Bricknell, R. H., Scripta Metall., 23 (1989), 599.

[14] Gabrielli, F., and Pelloux, R. M., Met. Trans., 13A (1982), 1083.

[15] Bain, K. R., and Pelloux, R. M., Proc. of Conf. on Superalloys, The Metall. Soc./AIME, Warrendale, PA (1984), 387.

[16] Bain, K. R., and Pelloux, R. M., Met. Trans. 15A (1984), 381.

[17] Andrieu, E., Molins, R., Ghomen, H., and Pineau, A., Mat. Sci. and Eng., 21 (1992), A154.

[18] Molins, R., Hochstetter, G., Chassaigne, J. C., and Andrieu, E., Acta Mater., 45, 2 (1997), 663.

[19] Lynch, S. P., et al., Fatigue Fract. Engr. Mater. Struct., 17 (1994), 297.

[20] Valerio, P., Gao, M., and Wei, R. P., Scripta Metall. Mater., 30, 10 (1994), 1269.

[21] Gao, M., Dwyer, D. J., and Wei, R. P., Superalloys 718, 625, 706 and Various Derivatives, E. A. Loria, ed., The Minerals, Metals and Minerals Society, Warrendale, PA, (1994), 581 pp.

[22] Gao, M., Dwyer, D. J., and Wei, R. P., Scripta Metall. Mater., 32, 8 (1995), 1169.

[23] Dwyer, D. J., Pang, X. J., Gao, M., and Wei, R. P., Applied Surface Science, 81 (1994), 229.

[24] Miller, C. F., Simmons, G. W., and Wei, R. P., Scripta Mater., 42 (2000), 227.

[25] King, B. R., Patel, H. C., Gulino, D. A., and Tatarchuk, B. J., Thin Solid Films, 192 (1990), 351.

[26] He, Thesis: Master of Science, "The Interaction of O_2 and H_2O with an Inconel 718 Surface," University of Maine (August 1996).

[27] Pang, X., Thesis: Master of Science, "Surface Study on Nickel Base Alloy Inconel 718 (001) Single Crystal," University of Maine (August 1993).

[28] Pang, X. J., Dwyer, D. J., Gao, M., Valerio, P., and Wei, R. P., Scripta Metall. Mater., 31, 3 (1994), 345.

[29] Gao, M., and Wei, R. P., Scripta Mater., 32, 7 (1995), 987.

[30] Gao, M., and Wei, R. P., Scripta Mater., 37, 12 (1997), 1843.

[31] Chen, S. F., and Wei, R. P., *Matls Sci. & Engr.*, A256 (1998), 197.

[32] Miller, C. F., Simmons, G. W., and Wei, R. P., Scripta Mater., 42 (2000), 227.

[33] Miller, C. F., Simmons, G. W., and Wei, R. P., Scripta Mater., 44 (2001), 2405.

[34] Miller, C. F., Simmons, G. W., and Wei, R. P., Scripta Mater., 48 (2003), 103.

[35] Miller, C. F., Ph.D. Dissertation, "Chemical Aspects of Environmentally Enhanced Crack Growth in Ni-Based Superalloys," Lehigh University (March 2001).

[36] Huang, Z. F., Iwashita, C., Chou, I., and Wei, R. P., "Environmentally Assisted Sustained Load Crack Growth in PM Nickel-Based Superalloys," Met. & Mater. Trans. A, 33A (2002), 1681.

[37] Huang, Z. F., "Oxygen Enhanced Crack Growth in Nickel-based Superalloys," Ph.D. Dissertation, Lehigh University (2002).

[38] Wei, R. P., *Advanced Technologies for Superalloy Affordability*, K. M. Chang, S. K.

[39] Srivastava, Furrer, D. U., and Bain, K. R., eds. The Minerals, Metals & Materials Society, Warrendale, PA (2000), 103.

[40] Nakajima, H., Nagata, S., Morishima, Y., Takahiro, H., and Yamaguchi, S., Defect and Diffusion Forum, 369 (1993), 95–98.

[41] Takada, J., Yamamoto, S., Kikuchi, S., and Adachi, M., Metall. Trans. A, 17A (1986), 221.

[42] Dowling, N. E., *Mechanical Behavior of Materials: Engineering Methods for Deformation, Fracture and Fatigue,* 2nd ed, Prentice Hall, Inc., Upper Saddle River, NJ (1999).

[43] Courtney, T. H., *Mechanical Behavior of Materials*, 2nd ed, McGraw-Hill Co., New York (2000).

[44] Atkinson, Taylor, M., and Hughes, A. E., Phil. Mag. A., 45 (1982), 823.

[45] Atkinson, M. L., Dwyer, and R., Taylor, J. Mater. Sci., 18 (1983), 2371.

[46] Atkinson, Solid State Ionics, 12 (1984), 309.

9 Subcritical Crack Growth: Environmentally Assisted Fatigue Crack Growth (or Corrosion Fatigue)

9.1 Overview

In Chapter 8, the essential framework and methodology for quantifying the influences of environment on crack growth was described. Here, environmentally assisted fatigue crack growth (or corrosion fatigue) in gaseous and aqueous environments, and its conjoint action with stress corrosion cracking, are considered. Illustrations (constrained by the "windows of opportunity" to a large extent) are drawn from research in the author's laboratory, and will highlight aluminum alloys, titanium alloys, and high-strength steels [1–15]. The approach follows that used for stress corrosion cracking, and focuses on coordinated experiments and analyses that probe the underlying chemical, mechanical, and materials interactions for crack growth. Linkage of the fracture mechanics based approach to the traditional stress-life (S-N) approach is made to provide a "common basis" for the interpretation and utilization for fatigue data in design, and to address "key (physically based) sources" for variability in S-N data. The various processes, and their inter-relationships, are depicted in the schematic diagrams shown previously in Fig. 8.7. Their incorporation into models for fatigue crack growth, however, is different, and is presented in Section 9.2.

It should be noted that, in corrosion fatigue, manifestations of environmental effects are reflected in a frequency dependence that gives rise to increase in fatigue crack growth rate (per cycle) with decreasing loading frequency that cannot be attributed to concomitant stress corrosion cracking. However, both sources for subcritical crack growth must be incorporated for operations (such as electrical power plants), in which the effects of both "on-off" and steady operating loads must be considered.

9.2 Modeling of Environmentally Enhanced Fatigue Crack Growth Response

The basic approach to the understanding and "prediction" of fatigue crack growth response is identical to that for stress corrosion cracking, except that the processes are treated on a cycle (*vis-à-vis*, time) basis, and resides in the following postulate and corollary as well, *i.e.*:

> **"Environmentally enhanced crack growth results from a <u>sequence</u> of processes and is <u>controlled</u> by the <u>slowest</u> process in the sequence."**

> **"Crack growth response reflects the dependence of the rate controlling process on the environmental, microstructural and loading variables."**

This fundamental hypothesis reflects the conceptual existence of an increment of fatigue crack growth, corresponding to the maximum for a given driving force ΔK, over which the growth rate is the weighted average of the environmentally affected and unaffected components; namely:

$$\left(\frac{da}{dN}\right)_e = \left(\frac{da}{dN}\right)_r \phi_r + \left(\frac{da}{dN}\right)_c \phi_c, \quad \text{where } \phi_r + \phi_c = 1 \qquad (9.1)$$

In Eqn. (9.1), $(da/dN)_e$ is the fatigue crack growth rate in an inert environment, and $(da/dN)_c$ is the maximum corrosion fatigue crack growth rate in the deleterious environment at the given ΔK level. The terms ϕ_r and ϕ_c are the areal fractions of deformation-controlled and environmentally affected (or pure and corrosion) fatigue crack growth rates, respectively. As such, when $\phi_c = 0$, $(da/dN)_e$ equals the "pure" fatigue rate $(da/dN)_r$, and when $\phi_c = 1$ ($\phi_r = 0$), $(da/dN)_e$ equals the full "corrosion fatigue" rate $(da/dN)_c$. (Keep in mind that Eqn. (9.1) reflects, on average, changes in crack front geometry, microstructure, and environmental conditions). Note here, the crack growth rate is "cycle-based," *vis-à-vis* "time-based" as in stress corrosion cracking, and requires an alternative procedure to account for the periodic nature of "chemical contributions" associated with cyclic loading.

If the material is susceptible to stress corrosion cracking (*i.e.*, environmentally affected cracking under sustained loads), a contribution from this mechanism (for applications such as power plant equipment) must be incorporated, and is treated as an additive term. As such:

$$\left(\frac{da}{dN}\right)_e = \left(\frac{da}{dN}\right)_{cyc} + \left(\frac{da}{dN}\right)_{tm}$$

where

$$\left(\frac{da}{dN}\right)_{cyc} = \left(\frac{da}{dN}\right)_r \phi_r + \left(\frac{da}{dN}\right)_c \phi_c \qquad (9.2)$$

$$\left(\frac{da}{dN}\right)_{tm} = \left(\int_0^\tau \left[\frac{da}{dt}(K(t))\right]_{cr} dt\right) \psi_{cr} + \left(\int_0^\tau \left[\frac{da}{dt}(K(t))\right]_{scc} dt\right) \psi_{scc}$$

$$\text{where } \phi_r + \phi_c = 1 \quad \text{and} \quad \psi_{cr} + \psi_{scc} = 1$$

In essence, for example, $(da/dN)_{cyc}$ might represent the power-on and power-off portion of a power plant's boiler duty cycle, and $(da/dN)_{tm}$ might represent, the steady-state portion of its duty cycle. It is recognized that the crack growth model implicitly requires continued *to-and-fro* adjustment in the local driving forces and cracking mechanisms to sustained orderly crack growth. This requirement is supported by experimental data gathered over the past four decades.

Linkage between corrosion fatigue crack growth response *per se* and the underlying chemical/electrochemical processes is established through the identification of the extent of surface reaction per cycle θ with the areal fraction by corrosion fatigue ϕ_c; namely,

$$
\begin{aligned}
\left(\frac{da}{dN}\right)_e &= \left(\frac{da}{dN}\right)_r + \left[\left(\frac{da}{dN}\right)_c - \left(\frac{da}{dN}\right)_r\right]\phi_c \\
&= \left(\frac{da}{dN}\right)_r + \left[\left(\frac{da}{dN}\right)_c - \left(\frac{da}{dN}\right)_r\right]\theta
\end{aligned}
\tag{9.3}
$$

where $(da/dN)_r$ is the fatigue crack growth rate in an "inert" reference environment, and $(da/dN)_c$ is the maximum crack growth rate in the deleterious environment of interest. The extent of surface reaction θ is identified with the period (or 1/frequency) of loading and the fraction of crack-tip surface that is undergoing fatigue. The identification of surface coverage θ with the areal fraction by corrosion fatigue ϕ_c is significant, and reflects the local distribution in fracture modes.

Specifically, in Eqn. (9.3), when $\theta = \phi_c = 0$ (*i.e.*, in an inert environment) the crack growth rate is equal that in an inert environment. Whereas, when $\theta = \phi_c = 1$, the growth rate reflects the full effect of the environment. The crack growth dependence on each of the rate-controlling processes is summarized herein and will be highlighted through specific examples. Note that Eqn. (9.3) and its derivatives reflect the functional dependences on the underlying rate-controlling process, but is not a "predictor" of the actual crack growth rates.

9.2.1 Transport-Controlled Fatigue Crack Growth

For highly reactive gases/active surfaces (*e.g.*, water vapor/aluminum), the rate of reaction of the environment with the newly created crack surfaces at the crack tip is limited by the rate of transport of gases by molecular (Knudsen) flow to the crack tip. The extent of reaction with the newly created crack surface, or surface coverage θ is proportional to the rate of arrival of the gas and the time for reaction as described in Chapter 8; namely,

$$
\theta \approx \frac{Fp_o}{SN_okT}t \propto \frac{p_o}{f}
\tag{9.4}
$$

In Eqn. (9.4), F is the volumetric flow rate, p_o is the external pressure of the deleterious gas, S is the surface area (both sides) of the crack increment, N_o is the density of surface sites (or number of metal atoms per unit area), k is Boltzmann's constant, and T is the absolute temperature. Note that, because of the transport by molecular (Knudsen) flow along the crack, the gas pressure at the crack tip would be orders of magnitude less than p_o.

By defining the fractional surface coverage θ as $\theta = (p_o/f)/(p_o/f)_s$, and substituting it into Eqn. (9.3), The functional dependence on vapor pressure and frequency becomes:

$$\left(\frac{da}{dN}\right)_e = \left(\frac{da}{dN}\right)_r + \left[\left(\frac{da}{dN}\right)_c - \left(\frac{da}{dN}\right)_r\right]\frac{(p_o/f)}{(p_o/f)_s} \tag{9.5}$$

where $(p_o/f)_s$ is the exposure needed to produce complete coverage of the freshly exposed crack surfaces during the loading cycle. The parameter p_o is the external pressure of the deleterious environment, and f is the frequency of loading. The fatigue crack growth rates $(da/dN)_e$, $(da/dN)_r$, and $(da/dN)_c$ are those for the given environment, an inert reference environment, and the "maximum" rate for the given environment at the specific crack-driving force ΔK level. If the temperature dependence is explicitly incorporated, Eqn. (9.5) then becomes:

$$\left(\frac{da}{dN}\right)_e = \left(\frac{da}{dN}\right)_r + \left[\left(\frac{da}{dN}\right)_c - \left(\frac{da}{dN}\right)_r\right]\frac{(p_o/fT^{1/2})}{(p_o/fT^{1/2})_s} \tag{9.6}$$

9.2.2 Surface/Electrochemical Reaction-Controlled Fatigue Crack Growth

Similar to the case of sustained loading, it is assumed that the surface reaction(s) that control crack growth follow first-order kinetics. As such the surface coverage θ of a deleterious gas is given by:

$$\frac{d\theta}{dt} = k(1-\theta); \quad k = k_o \exp\left(-\frac{E_S}{RT}\right)$$

and

$$\theta = 1 - \exp(-kt) = 1 - \exp\left[-\frac{k_o}{f}\exp\left(-\frac{E_S}{RT}\right)\right] \tag{9.7}$$

where k and k_o are reaction rate constants, E_S is the activation energy, R is the universal gas constant, and T is the temperature.

For electrochemical reaction-controlled crack growth, the reactions are reflected through the reaction current density i, where:

$$i = i_o \exp(-kt); \quad q = \frac{i_o}{k}[1 - \exp(-kt)]$$

and

$$\theta = \frac{q}{q_o} = 1 - \exp\left[-\frac{k_o}{f}\exp\left(-\frac{E_{ec}}{RT}\right)\right] \tag{9.8}$$

Here, k and k_o are reaction rate constants, and E_{ec} is the activation energy for electrochemical reaction, R is the universal gas constant, and T is the temperature.

Substitution of Eqns. (9.7) and (9.8) into Eqn. (9.3) yields Eqn. (9.9) for gases, and Eqn. (9.10) for aqueous environments:

$$\left(\frac{da}{dN}\right)_e = \left(\frac{da}{dN}\right)_r + \left[\left(\frac{da}{dN}\right)_c - \left(\frac{da}{dN}\right)_r\right]\left\{1 - \exp\left[-\frac{k_o}{f}\exp\left(-\frac{E_S}{RT}\right)\right]\right\} \tag{9.9}$$

and

$$\left(\frac{da}{dN}\right)_e = \left(\frac{da}{dN}\right)_r + \left[\left(\frac{da}{dN}\right)_c - \left(\frac{da}{dN}\right)_r\right]\left\{1 - \exp\left[-\frac{k_o}{f}\exp\left(-\frac{E_{ec}}{RT}\right)\right]\right\} \quad (9.10)$$

Needless to say, the reaction rate constants k_o and the activation energies must be appropriate for the given material-environment system.

9.2.3 Diffusion-Controlled Fatigue Crack Growth

If the transport and surface reaction processes are fast (*i.e.*, not rate limiting), then crack growth would be controlled by the rate of diffusion of the embrittling species into the fracture process zone ahead of the crack tip. The functional dependence for diffusion-controlled crack growth, therefore, assumes the following form:

$$\frac{da}{dN} \propto \frac{p_o^m}{f^{1/2}}\exp\left(-\frac{E_D}{2RT}\right) \quad (9.11)$$

The exponent m in Eqn. (9.11) is typically assumed to be equal to $1/2$ for diatomic gases, such as hydrogen; but the number m is used here to recognize the possible existence of intermediate states in the dissociation from their molecular to atomic form. The factor 2 in the exponential term again recognizes the dissociation of diatomic gases, such as hydrogen (H_2), into atomic form.

9.2.4 Implications for Material/Response

Note that the functional dependence of fatigue crack growth response can be quite different between the different material-environment systems. These differences arise from differences in reactivity, mechanisms, kinetics, etc., and must be characterized carefully.

9.2.5 Corrosion Fatigue in Binary Gas Mixtures [3]

The foregoing models for corrosion fatigue crack growth have been extended to the consideration of crack growth in gas mixtures [3]. For simplicity, the case of binary gas mixture was considered, in which one of the component gases was taken to be an inhibitor (*i.e.*, a gas that would react with the clean metal surface, thus "blocking" reaction sites, but it would not produce enhancement in crack growth). The model is important for examining, for example, the influence of oxygen (acting as an inhibitor) on fatigue crack growth in moist (humid) air, where water vapor acts as the damaging species.

   It is assumed that (i) both gases are strongly adsorbed on the clean metal surfaces produced by cracking, (ii) chemical adsorption of either gas at a given surface site would preclude further adsorption at that site, (iii) the ratio of partial pressures of the gases at the crack tip is essentially the same as that of the surrounding (external) environment, (iv) no capillary condensation of either gas occurs at the crack tip, and (v) there are no reactions between the two gases to form new phases. In accordance with Weir *et al.* [3], the cycle-dependent component of crack growth rate in the gas mixture, $(da/dN)_{cf,m}$, is assumed to be proportional to the extent of surface reaction with the deleterious gas during one loading cycle (θ_a). Assuming first-order reaction kinetics for both gases with respect to pressure and available surface sites, the reaction rates are given as follows:

$$\frac{d\theta_a}{dt} = k_a p_a (1 - \theta)$$

$$\frac{d\theta_i}{dt} = k_i p_i (1 - \theta)$$

(9.12)

The subscripts a and i denote the deleterious and inhibitor gases, respectively. The quantities k_a, p_a, k_i, and p_i are, respectively, the reaction rate constants and partial pressures of the gases at the crack tip. The coverages θ_a and θ_i denote the fraction of crack-tip surface that has reacted with the deleterious and inhibitor gases, respectively, with the total coverage $\theta = \theta_a + \theta_i$ and $0 \le \theta \le 1$.

Equation (9.12) may be solved straightforwardly to obtain the extent of reaction, or surface coverage, with each gas as follows:

$$\theta_a = \frac{k_a p_a}{k_a p_a + k_i p_i} \{1 - \exp[-(k_a p_a + k_i p_i) t]\}$$

$$\theta_i = \frac{k_i p_i}{k_a p_a + k_i p_i} \{1 - \exp[-(k_a p_a + k_i p_i) t]\}$$

(9.13)

The surface coverage by the deleterious gas (θ_a) relative to the total surface coverage is given by solving Eqn. (9.13), and is given by Eqn. (9.14) [3]:

$$\frac{\theta_a}{\theta} = \frac{\theta_a}{(\theta_a + \theta_i)} = \left[1 + \frac{k_i p_i}{k_a p_a}\right]^{-1}; \quad 0 \le \theta \le 1$$

(9.14)

If the combination of total pressure of the gas mixture at the crack tip $(p_m = p_a + p_i)$ and the cyclic loading frequency (f), namely, p_m/f, is such that the reactions with the newly exposed crack surface are completed, then $\theta = 1$ and θ_a achieves its maximum value θ_{am}; namely,

$$\theta_{am} = \left[1 + \frac{k_i p_i}{k_a p_a}\right]^{-1}; \quad (\text{for } \theta = 1)$$

(9.15)

Because the corrosion fatigue crack growth rate is proportional to θ_a, the maximum cycle-dependent term at a given ΔK level is given in terms of the growth rates in pure (deleterious) gas and in the inert (reference) environment, along with θ_{am}, as follows:

$$\left(\frac{da}{dN}\right)_{cf,m} = \left[\left(\frac{da}{dN}\right)_c - \left(\frac{da}{dN}\right)_r\right]\left[1 + \frac{k_i p_i}{k_a p_a}\right]^{-1}; \quad (\text{for } \theta = 1)$$

(9.16)

In Eqn. (9.16), the partial pressures refer to those at the crack tip. They may be replaced by the partial pressures in the external environment if the relative pressure attenuation along the crack is the same for both gases. Because only the competitive adsorption between the two gases was modeled, the attenuation in rates between the two component gases would apply equally well to transport, surface reaction, and diffusion-controlled crack growth.

9.2.6 Summary Comments

The foregoing models provide the essential link between the fracture mechanics and surface chemistry/electrochemistry aspects of fatigue crack growth response. Crack growth response, in fact, is the response of a material's microstructure to the conjoint actions of the mechanical and chemical driving forces. In the following sections, the responses in gaseous and aqueous environments are illustrated through selected examples from the works of the author and his colleagues (faculty, researchers, and graduate students) over past years.

9.3 Moisture-Enhanced Fatigue Crack Growth in Aluminum Alloys [1, 2, 5]

Fracture mechanics and surface chemistry studies were carried out to develop a clearer understanding of the enhancement of fatigue crack growth by deleterious, gaseous environments. These studies were complemented by fractographic examinations to gain understanding of the alloy's microstructural response. Here, a comprehensive study of moisture-enhanced fatigue crack growth in a 2219-T851 (AlCu) aluminum alloy is summarized. Study of a 7075-T651 (AlMgZn) aluminum alloy is summarized to affirm and enhance this broad-based understanding.

9.3.1 Alloy 2219-T851 in Water Vapor [1, 2]

Data on the influence of (pure) water vapor, at pressures from 1 to 26.6 Pa, on the kinetics of fatigue crack growth (*i.e.*, (da/dN) versus ΔK) at room temperature, are shown in Fig. 9.1, along with data obtained in dehumidified argon. The data at 26.6 Pa are comparable with those obtained in air (at 40 to 60 percent relative humidity), distilled water, and 3.5 percent NaCl solution [1]. The data in dehumidified argon correspond to those in vacuum at less than 0.50 μPa. These data are also shown in Fig. 9.2 as a function of water vapor pressure at three ΔK levels. The error bands represent ninety-five percent confidence intervals computed from the residual standard deviations in each set of data. The results in Fig. 9.2 show that at a frequency of 5 Hz, the rate of crack growth is essentially unaffected by water vapor until a threshold pressure is reached. (This threshold pressure is attributable to the significant transport limitation at these very low water vapor pressures.) The rate then increased and reached a maximum within one order of magnitude increase in vapor pressure from this threshold. The maximum rate is equal to that obtained in air, distilled water, and 3.5 percent NaCl solution (at 20 Hz). The transition range, in terms of pressure/frequency, is comparable to that reported by Bradshaw and Wheeler [9] on another aluminum alloy.

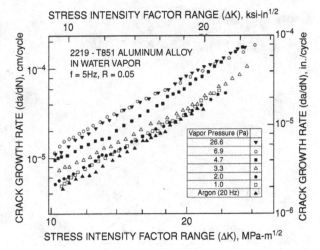

Figure 9.1. Influence of water vapor pressure on the kinetics of fatigue crack growth in 2219-T851 aluminum alloy at room temperature [2].

Representative scanning electron microscopy (SEM) microfractographs of a specimen tested in water vapor at 4.66 Pa (*i.e.*, within the transition region from 0 to 8 Pa in Fig. 9.2) are shown in Fig. 9.3, and are compared with those taken from specimens, one tested in dehumidified argon and the other in water vapor at 26.6 Pa (*i.e.*, one reflecting full environmental effect and the other no environmental effect) (Fig. 9.4). The microfractographs clearly show differences in fracture surface morphology. It is seen that the fracture surface morphology in the mid-thickness region (Fig. 9.3(a)) is comparable with that associated with crack growth in dehumidified argon (Fig. 9.4(a)). The fracture surface morphology in the near-surface region,

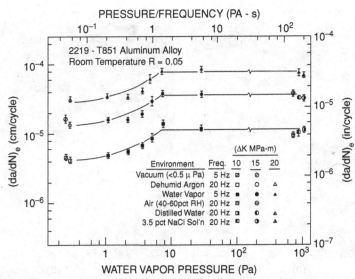

Figure 9.2. Influence of water vapor pressure (or pressure/frequency) on fatigue crack growth rates in 2219-T851 aluminum alloy at room temperature. Solid line represents model predictions [2].

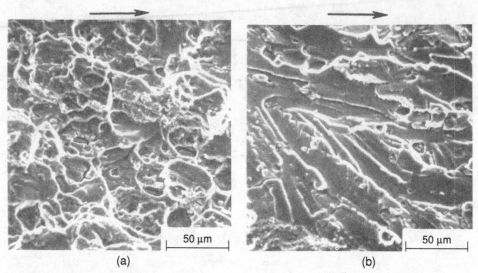

(a)   (b)

Figure 9.3. SEM micrographs taken from the mid-thickness region (center) (a) and the near-surface region (edge) (b) of the specimen showing differences in surface morphology ($\Delta K =$ 16.5 MPa-m$^{1/2}$, $R = 0.05$, $f = 5$ Hz, and 4.06 Pa H_2O Vapor) [2].

on the other hand, corresponds to that associated with crack growth in humidified argon (at 26.6 Pa H_2O vapor), Fig. 9.3(b) versus Fig. 9.4(b), and reflects full effect of the water vapor.

The reactions of water vapor with clean surfaces of 2219-T851 aluminum alloys were studied by Auger electron spectrometry (AES) and x-ray photoelectron spectroscopy (XPS) and are presented in [2]. Changes in the normalized oxygen Auger (510 eV) signal as a function of exposure to water vapor are shown in Fig. 9.5.

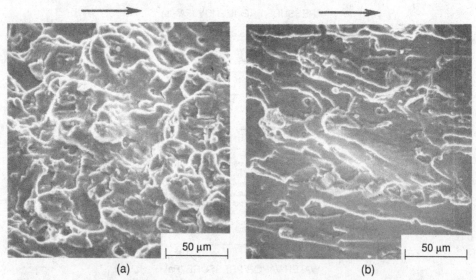

(a)   (b)

Figure 9.4. SEM micrographs of specimens tested in argon and in water vapor at 26.6 Pa (full environmental effect) showing similar differences in fracture surface morphology as seen in Fig. 9.3: (a) argon, (b) 26.6 Pa H_2O vapor. ($\Delta K = 16.5$ MPa-m$^{1/2}$, $R = 0.05$, $f = 5$ Hz) [2].

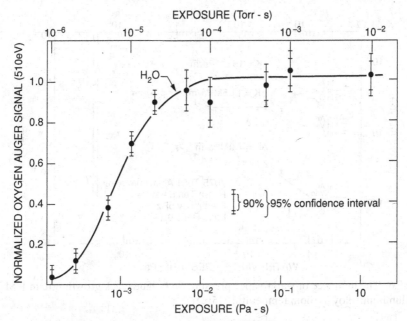

Figure 9.5. Kinetics of reactions of water vapor with 2219-T851 aluminum alloy at room temperature [2].

Normalization is based on the average value of oxygen Auger (510 eV) signals from specimens exposed to water vapor for 6.65×10^{-2} to 1.33 Pa-s. Comparable results were obtained from the companion XPS studies. The results show that the initial rate of reaction of clean aluminum surfaces with water vapor is rapid and reaches "saturation" after about 2.7×10^{-3} Pa-s exposure; that is, the extent of reaction with aluminum is limited. XPS results indicate that the reactions are associated with the formation of an oxide or a hydrated oxide layer. The limited reactions with water vapor are consistent with previous results on a high-strength AISI 4340 steel [4]. The rate of reaction, however, is 10^8 to 10^9 times faster than the corresponding rate (associated with the slow, second step) of reaction with AISI 4340 steel.

9.3.2 Alloy 7075-T651 in Water Vapor and Water [5]

To further understand the influence of environment on fatigue crack growth, the responses of a 7075-T651 (AlMgZn) alloy to changes in water vapor pressure, at room temperature, and a test frequency of 5 Hz, is shown in Fig. 9.6 [5]. With increasing water vapor pressure (from about 1 Pa), the rate of crack growth increased and reached an intermediate plateau at about 5 Pa. Above about 70 Pa, there were further increases in growth rates with increasing pressure, with a maximum equal with those attained in water. The fatigue crack growth rates in oxygen are comparable with those observed at the very low water vapor pressures, while the rates in vacuum (10^{-6} Pa) and in dehumidified argon were somewhat higher.

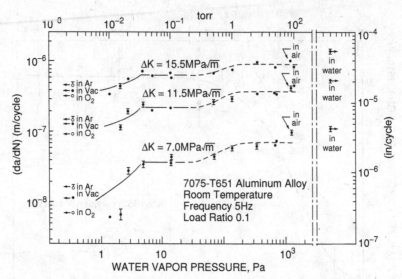

Figure 9.6. The influence of water vapor pressure on fatigue crack growth rate in I/M 7075-T651 aluminum alloy at room temperature [5].

The changes in crack growth rates with water vapor exposure (pressure/frequency) appear to be essentially independent of the stress intensity (ΔK) level. The observed response is consistent with other aluminum (AlCu, AlCuMg, and AlMgZn) alloys, except that the increase in rates above the first plateau (*cf.* 2219 and 7075 alloys) appear to be limited to the Mg-containing alloys, and is attributed to the reactions of water vapor with magnesium in the alloy and the resulting, further, production of hydrogen [5].

9.3.3 Key Findings and Observations

The principle findings are as follows: (a) The reaction of water vapor with aluminum is very rapid, and results in the formation of oxides or hydrated oxides. What needs to be recognized is that these oxidation reactions are accompanied by the release of hydrogen, which might be the real "trouble maker." (b) These reactions are very rapid, and are completed at exposures on the order of 10^{-4} Pa-s, compared with about 1 Pa-s of equivalent "exposure" to attain "saturation" in fatigue crack growth rates. (c) For water vapor, there is no evidence for metal dissolution. Two major points need to be recognized. First, the observed four orders of magnitude difference between the fatigue-cracking response and surface reaction kinetics support the identification of "transport control" of crack growth. Second, the evidence tends to support "hydrogen embrittlement" as the mechanism for the enhancement of crack growth. (The fact that hydrogen does not dissociate directly on aluminum precludes a direct validation of this mechanism.)

The observed response is a function of frequency and temperature [1, 5]. In reality, the dependence is on pressure/frequency, or on the exposure (pressure × time) to the environment. With increasing temperature, the exposure (pressure/frequency) required to reach the plateau rate would decrease, and reflects the increased rate of reaction with the metal surfaces. The crack growth rate itself also reflects the deformation response of the alloys and strongly depends on temperature. Crack growth enhancement also depends on material thickness, load ratio (R), and ΔK level; their influences need to be fully explored for structural integrity and durability.

9.4 Environmentally Enhanced Fatigue Crack Growth in Titanium Alloys [6]

Parallel fracture mechanics and surface reaction and surface chemistry studies were carried out to develop understanding of environmentally assisted crack growth in titanium alloys [6]. Room temperature crack growth response in water vapor was determined for annealed Ti-5Al-2.5Sn alloy and Ti-6Al-4V alloy in the solution-treated and solution-treated plus overaged conditions as a function of water vapor pressure from 0.266 to 665 Pa at a frequency of 5 Hz and a load ratio R of 0.1. The results are compared with data obtained in vacuum. The kinetics of reactions of water vapor and oxygen with fresh surfaces of these alloys were measured by Auger electron spectroscopy (AES) at room temperature. The results of limited additional studies on the influences of loading frequency and temperature are included to highlight the unanticipated influences of strain/strain-rate-induced hydride formation on fatigue crack growth.

9.4.1 Influence of Water Vapor Pressure on Fatigue Crack Growth

The influence of water vapor pressure on the kinetics of fatigue crack growth (at $R = 0.1$ and $f = 5$ Hz) in Ti-6Al-4V alloy in the solution-treated (ST) and solution-treated and overaged (STOA) conditions were examined at room temperature, in vacuum (below 7×10^{-7} Pa) and in pure water vapor pressures from 0.266 to 665 Pa [6]. Limited fatigue crack growth experiments were carried out also on an annealed Ti-5%Al-2.3Sn alloy in vacuum and in water vapor at 133 Pa to provide direct linkage to the surface reaction data for water vapor and oxygen. The results for Ti-6Al-4V in the ST and STOA condition are shown in Figs. 9.7 and 9.8, respectively. Those for the Ti-5Al-2.5Sn are shown in Fig. 9.9. The results on the Ti-6Al-4V alloy (compare Figs 9.7 and 9.8) suggest that saturation (*i.e.*, a maximum) in environmental effect had occurred at water vapor pressure above about 25 Pa.

9.4.2 Surface Reaction Kinetics

The kinetics of reactions of water vapor and oxygen with titanium alloy surfaces at room temperature were determine by AES [6]. The measurements were limited to the Ti-5Al-2.5Sn alloy, and reflected principally the reactions of titanium with these gases, and are deemed to be applicable to the Ti-6Al-4V alloys as well.

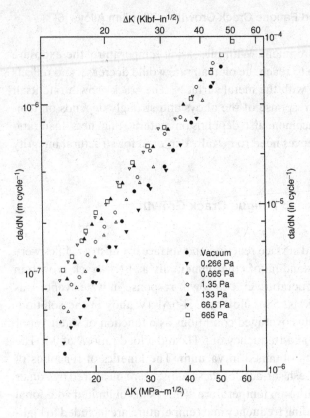

Figure 9.7. Influence of water vapor pressure on the kinetics of fatigue crack growth in solution-treated Ti-6Al-4V ally at room temperature ($R = 0.1$, $f = 5$ Hz) [6].

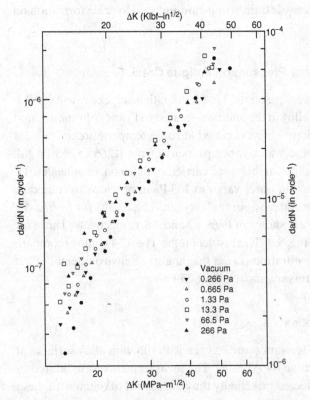

Figure 9.8. Influence of water vapor pressure on the kinetics of fatigue crack growth in solution-treated plus overaged Ti-6Al-4V alloy at room temperature ($R = 0.1$, $f = 5$ Hz) [6].

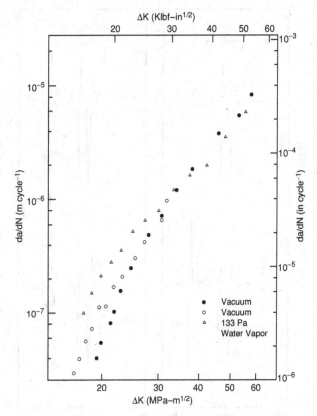

Figure 9.9. Influence of water vapor pressure on the kinetics of fatigue crack growth in Ti-5Al-2.5Sn alloy at room temperature ($R = 0.01$, $f = 5$ Hz) [6].

Auger electron spectra of Ti-5Al-2.5Sn surfaces are shown in Fig. 9.10. Only the signals for the alloying elements (Ti and Sn) are shown, along with that of oxygen. The signal for aluminum at 1400 eV is not included. Spectrum (a) shows the composition of the "clean" surface exposed by impact fracture in vacuum; spectrum (b) after exposure to water vapor for 5.3×10^{-3} Pa-s; (c) after exposure to oxygen for 5.3×10^{-3} Pa-s; and (d) after exposure to water vapor at 1.33 kPa. The oxygen uptake that is associated with exposures to water vapor and to oxygen is shown in Figs. 9.11 and 9.12, respectively.

9.4.3 Transport Control of Fatigue Crack Growth

From the surface chemistry studies, it is seen that the titanium-water vapor surface reaction rate is very fast. The reaction rate constant k_c at room temperature is of the order of 10^3 Pa^{-1}s^{-1}, and supports the transport controlled for crack growth. (Although there is a slower additional reaction, its contribution to crack growth appears to be small, and no further consideration was given to it.) Conformance of the data to the model for transport controlled crack growth is shown in Fig. 9.13 at two ΔK levels and two heat treatment conditions.

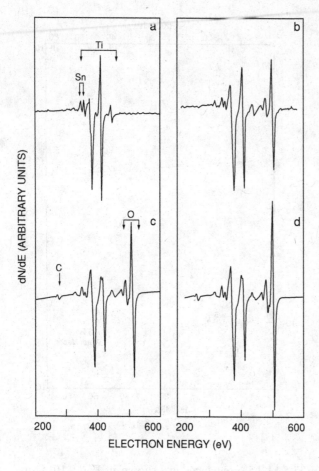

Figure 9.10. Auger election spectra of Ti-5Al-2.5Sn surfaces: spectrum a, after impact fracture in vacuum; spectrum b, after exposure to 5.3×10^{-3} Pa-s (4×10^{-5} torr-s) water vapor; spectrum c, after exposure to 5.3×10^{-3} Pa-s (4×10^{-5} torr-s) oxygen; spectrum d, after exposure to water vapor at 1.33 kPa (10 torr) ($E_p = 2$ KeV, 3 eV peak-to-peak, $I_p = 20$ μA) [6].

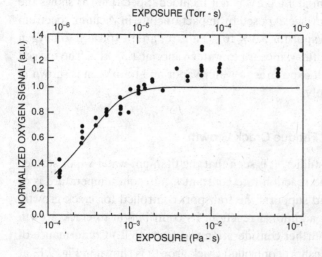

Figure 9.11. Normalized oxygen Auger electron signal versus water vapor exposure for Ti-5Al-2.5Sn alloy at room temperature [6].

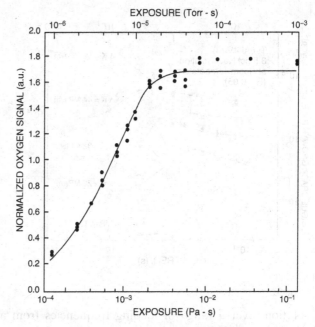

Figure 9.12. Normalized oxygen Auger electron signal versus oxygen exposure for Ti-5Al-2.5Sn alloy at room temperature [6].

9.4.4 Hydride Formation and Strain Rate Effects

For hydride-forming alloys, such as titanium alloys, the crack growth response may exhibit strong temperature and frequency dependence that is also a function of K level. This dependence reflects the influence of strain, strain rate, and temperature on hydride formation and rupture [7, 8]. Support for this response is provided by the early fatigue crack growth data on Ti-6Al-4V(Ti64) alloy in 0.6 M NaCl

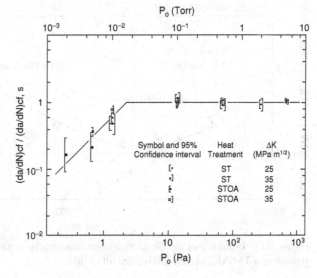

Figure 9.13. Comparison of normalized (corrosion) fatigue crack growth rates for solution-treated (ST) and solution-treated plus averages (STOA) Ti-6Al-4V alloy in water vapor with model predictions for pressure dependence at room temperature ($R = 0.1$, $f = 5$ Hz) [6].

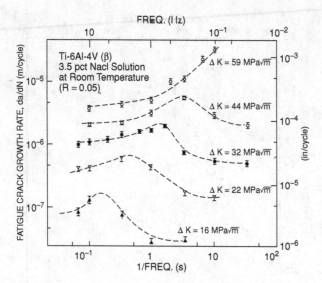

Figure 9.14. Influence of frequency on fatigue crack growth in a Ti-6Al-4V alloy in 0.6 M NaCl solution that reflects the propensity for hydride formation at the higher frequencies [7].

solution, over a range of loading frequencies from about 2×10^{-3} to 10 Hz at R = 0.1 (Fig. 9.14) [7], and by the temperature dependence of a Ti-6Al-2Sn-4Zr-2Mo-0.1S (Ti6242S) alloy at two levels of internal hydrogen (Fig. 9.15) [8].

For the Ti-6Al-4V alloy, the fatigue crack growth rate response in the higher-frequency domain, exhibiting a $p^{1/2}$-power dependence, the response is consistent with diffusion control. Furthermore, it suggests the formation of strain-induced hydrides and the concomitant increase in growth rate (Fig. 9.14). At specific combinations of frequency and ΔK, the fatigue crack growth rates decrease with further

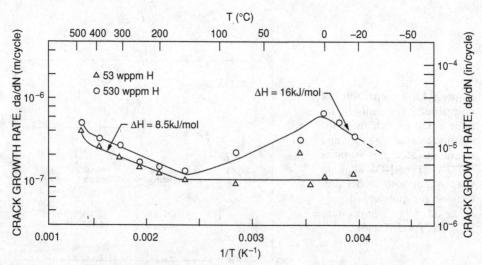

Figure 9.15. Influence of dissolved hydrogen concentration and temperature on fatigue crack growth in a Ti-6Al-2Sn-4Zr-2Mo-0.2Si alloy [8].

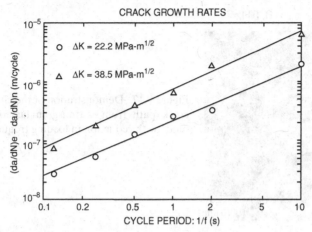

Figure 9.16. Variation of cycle-dependent component of corrosion fatigue crack growth rate with inverse frequency (or loading period) for an AISI 4340 steel tested in water vapor (at 585 Pa) at room temperature [10].

decrease in frequency (or strain rate), and is consistent with the decreased ability for the formation of hydrides. For the Ti6242S alloy, the influence of temperature and dissolved hydrogen concentration on hydride formation and hydrogen enhancement of crack growth is shown in Fig. 9.15. At temperatures above about 400 K the effect of hydrogen essentially disappears. This response can reflect control by hydrogen diffusion, but needs to be confirmed.

9.5 Microstructural Considerations

The role of micromechanisms (or of microstructure) is explicitly incorporated in the superposition model for crack growth through the formal identification of the areal fraction of surface area of the crack that undergo environmentally assisted cracking (ϕ) with the fraction (θ) that undergoes reaction with the environment. The implications of the model are the following: (i) the partitioning of hydrogen (and other deleterious gases) to the various microstructural sites would not be uniform, and (ii) the fractional area of the fracture surface (ϕ) produced by corrosion fatigue would be equal to the fractional surface area (θ) for chemical reactions (or deleteriously affected by the environment). The cycle-dependent component of corrosion fatigue crack growth rate, for AISI 4340 steel tested in water vapor (at 585 Pa) at room temperature, is shown as a function of inverse frequency (or period) in Fig. 9.16. Fractographic results show a change in fracture surface morphology, with decreasing frequency from 8 to 0.1 Hz, from a predominantly transgranular mode (relative to the prior austenite grains) to one that is predominantly intergranular (see Fig. 9.17). Figure 9.18 shows a composite that captures the change in fracture surface morphology, from transgranular to intergranular failure, with decreasing frequency from 8 to 0.1 Hz.

8 Hz   0.1 Hz

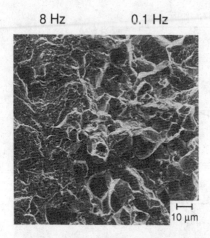

Figure 9.17. Demonstration of the change in fatigue crack path from a transgranular to intergranular mode with reduction in loading frequency. [10]

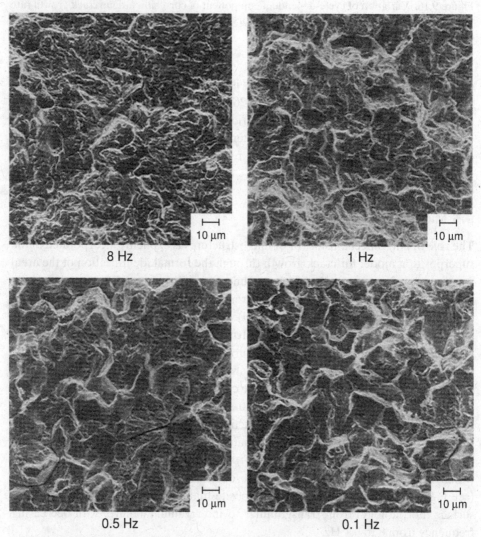

Figure 9.18. Demonstration of the gradual change in crack paths (transgranular to intergranular mode) with loading frequency [10].

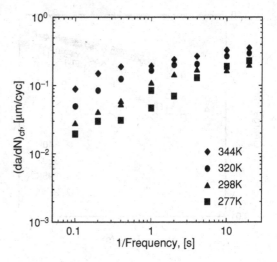

Figure 9.19. Demonstration of the temperature and inverse frequency dependence of crack growth rates in HY130 steel in 3.5 percent NaCl solution at different temperatures [11].

9.6 Electrochemical Reaction-Controlled Fatigue Crack Growth

For electrochemical reaction controlled fatigue crack growth, the dependence is explicitly given by Eqn. (9.10); specifically, by the second part; namely,

$$\left(\frac{da}{dN}\right)_{cf} = \left[\left(\frac{da}{dN}\right)_c - \left(\frac{da}{dN}\right)_r\right]\left\{1 - \exp\left[-\frac{k_o}{f}\exp\left(-\frac{E_{ec}}{RT}\right)\right]\right\}$$

This control is demonstrated through the direct comparisons of the inverse frequency dependence of fatigue crack growth rates at different temperatures. (The exponential term reflects the time, or 1/frequency, and temperature dependence of electrochemical reactions, including hydrogen production, with the newly created metal surfaces, at the crack tip.) This comparison was made in [11] and is summarized here. Fatigue crack growth measurements were conducted on a HY130 steel at a constant $\Delta K = 40$ MPa-m$^{1/2}$, in 3.5 percent NaCl solution at 277, 298, 320, and 344 K, over a range of loading frequencies from 0.05 to 10 Hz. The corrosion fatigue crack growth component, $(da/dN)_{cf}$, at 344, 320, 298, and 277 K, is plotted against 1/frequency in Fig. 9.19. The corresponding charge transfer versus time data, obtained from *in situ* fracture of notched round specimens in the same electrolyte, for the same temperatures, are shown in Fig. 9.20.

A comparison between the corrosion fatigue crack growth response (*i.e.*, $(da/dN)_{cf}$ versus $1/f$) and the electrochemical reaction (charge transfer) response (*i.e.*, q versus t) is made by matching the two sets of "independently obtained" data in Fig. 9.21. The "excellent" agreement between the fatigue crack growth and charge transfer (electrochemical reaction) data confirms the electrochemical reaction control of fatigue crack growth. The environmental influence is manifested in the frequency dependence. The predicted frequency and temperature dependence is affirmed by the result on a HY130 steel in an acetate buffer solution (pH 4.2) at the same temperatures, over frequencies from 0.1 to 10 Hz (Fig. 9.22) [12].

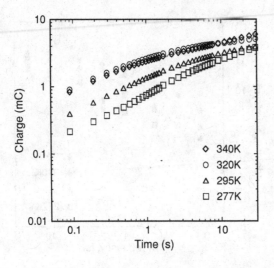

Figure 9.20. Corresponding (independently measured) charge transfer data, obtained from *in situ* fracture of notched round specimens of HY130 steel in 3.5 percent NaCl solution at the same temperatures [11].

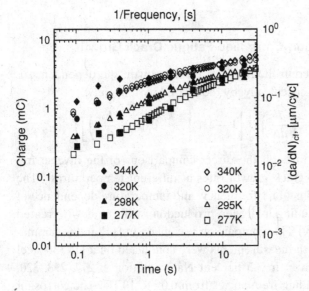

Figure 9.21. Direct confirmation of electrochemical reaction control of CF crack growth through correlation between the reaction and crack growth data [11].

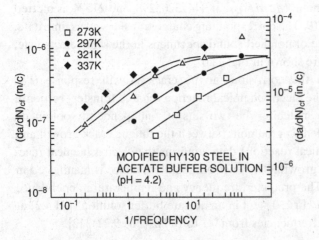

Figure 9.22. Affirmation of electrochemical reaction control by supporting data in an acetate buffer solution at the same temperatures [12].

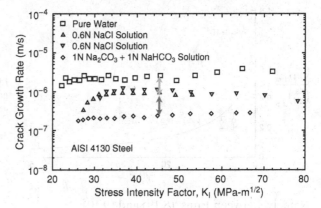

Figure 9.23. Manifestation of electrochemical reaction rate control through changes in crack growth rate with ionic species [13].

It is interesting to note that the influences of electrochemical/chemical reactions on crack growth response for fatigue and static loading are manifested differently. This difference is illustrated in Figs. 9.23 and 9.24. Figure 9.23 shows that, for sustained loading, or stress corrosion cracking in an AISI 4130 steel, anions in solution compete with the hydroxyl ions (from water dissociation) for adsorption sites, which leads to a three-fold reduction in stage II crack growth rates from distilled water to sodium chloride and carbonate-bicarbonate solutions [13]. Retardation in corrosion fatigue crack growth response (for an X-70 steel), on the other hand, is now manifested in the frequency domain (see Fig. 9.24) at a given crack growth rate [14]. Full environmental effect appear to be achievable; *albeit*, at much lower frequencies for the less reactive solutions.

The cause for this difference may be seen clearly through a comparison of the functional dependence on reaction rate constant between sustained-load (stress corrosion cracking, SCC) and fatigue (corrosion fatigue, CF) crack growth.

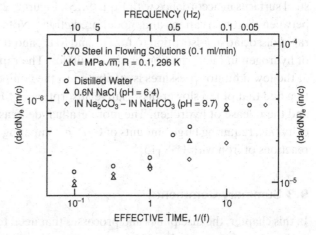

Figure 9.24. Manifestation of electrochemical reaction control changes in frequency dependence with ionic species [14].

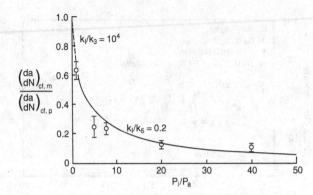

Figure 9.25. Influence of binary gas mixtures on fatigue crack growth [15].

Namely, between Eqns. (8.19) and (9.10):

$$\left(\frac{da}{dt}\right)_{II} \propto k_o \exp\left(-\frac{E_{ec}}{RT}\right)$$

$$\left(\frac{da}{dN}\right)_{cf} = \left[\left(\frac{da}{dN}\right)_c - \left(\frac{da}{dN}\right)_r\right]\left\{1 - \exp\left[-\frac{k_o}{f}\exp\left(-\frac{E_{ec}}{RT}\right)\right]\right\}$$

The responses reflect the respective functional dependence on the reaction rate constant k_o. For SCC, decreases in k_o lead to direct reductions in Stage II crack growth rates (see Fig. 9.23). For CF, on the other hand, because the dependence on k_o is embedded in the exponential term, it change is reflected in the l/frequency domain (see Fig. 9.24).

9.7 Crack Growth Response in Binary Gas Mixtures

The inhibiting effect of carbon monoxide (CO) on fatigue crack growth in a 2–1/4Cr-1 Mo steel exposed to H_2S/CO mixtures (such as that produced during coal gasification) is manifested in the competition for surface adsorption sites. It is expected that the embrittling impact of hydrogen that is released by the reactions with H_2S would be moderated by the competitive adsorption of CO on the clean steel surfaces in accordance with Eqn. (9.15). Figure 9.25 shows the good agreement between the observations and model predictions. Note that the reactions involve a rapid dissociative adsorption to for H^+ and SH^-, and the subsequent slower release of hydrogen in the formation of metal sulfides. The rapid decrease in relative rates at the low inhibitor pressures is attributed to the competition between CO adsorption and that of the slower dissociative adsorption of SH^- in forming metal sulfides and the release of hydrogen. The more gradual decrease reflected the greater difficulty (*i.e.*, requiring large amounts of CO) in competing against the very rapid initial reactions of iron with H_2S [15].

9.8 Summary Comments

In this chapter, the disciplines and processes that need to be brought to bear to solve *real* problems that involve fatigue cracking of materials in *realistic* environments are highlighted through selected examples. The development and use of tools for design and management of engineered systems, similar to that for stress corrosion cracking, must incorporate mechanistically based understanding and modeling of material response in terms of loading (stress analyses), environmental and microstructural variables. The foregoing discussions have dealt with fatigue cracking (crack growth) from a perspective based on linear elastic fracture mechanics. It is recognized that many materials can undergo corrosion damage (both in and out of service) and hasten crack nucleation and significantly shorten service life. Such interactions are discussed in brief here and are introduced in Chapter 10 in relation to probabilistic considerations.

REFERENCES

[1] Wei, R. P., and Simmons, G. W., "Recent Progress in Understanding Environment Assisted Fatigue Crack Growth," Int'l J. of Fracture, 17, 2 (1981), 235–247.

[2] Wei, R. P., Pao, P. S., Hart, R. G., Weir, T. W., and Simmons, G. W., "Fracture Mechanics and Surface Chemistry Studies of Fatigue Crack Growth in an Aluminum Alloy," Metallurgical Transactions A, 11A (1980), 151–158.

[3] Weir, T. W., Simmons, G. W., Hart, R. G., and Wei, R. P., "A Model for Surface Reaction and Transport Controlled Fatigue Crack Growth," ScriptaMet., 14 (1980), 357–364.

[4] Simmons, G. W., Pao, P. S., and Wei, R. P., "Fracture Mechanics and Surface Chemistry Studies of Subcritical Crack Growth in AISI 4340 Steel," Metallurgical Transactions A, 9A (1978), 1147–1158.

[5] Gao, M., Pao, P. S., and Wei, R. P., "Chemical and Metallurgical Aspects of Environmentally Assisted Fatigue Crack Growth in 7075–7651 Aluminum Alloy," Met. Trans. A, 19A (1988), 1739–1750.

[6] Gao, S. J., Simmons, G. W., and Wei, R. P., "Fatigue Crack Growth and Surface Reactions For Titanium Alloys Exposed to Water Vapor," Mat'ls. Sci. & Eng'g., 62, (1984), 65–78.

[7] Chiou, S., and Wei, R. P., "Corrosion Fatigue Cracking Response of Beta Annealed Ti-6Al-4V Alloy in 3.5% NaCl Solution," Report No. NADC-83126-60 (Vol. V), U. S. Naval Air Development Center, Warminster, PA (30 June 1984).

[8] Pao, P. S., and Wei, R. P., "Hydrogen-Enhanced Fatigue Crack Growth in Ti-6Al-2Sn-4Zr-2Mo-0.1Si," in Titanium: Science and Technology, G. Lutjering, U. Zwicker, and W. Bank, eds., FRG: Deutsche Gesellshaft für Metallkunde e.v. (1985), 2503.

[9] Bradshaw, F. J., and Wheeler, C., "The Effect of Environment on Fatigue Crack Growth in Aluminum and Some Aluminum Alloys," Applied Materials Research, 5 (1966), 112–120.

[10] Wei, R. P., and Gao, M., "Hydrogen Embrittlement and Environmentally Assisted Crack Growth," Hydrogen Effects on Material Behavior, N. R. Moody and A. W. Thompson, eds., The Mineral, Metals & Materials Society, Warrendale, PA (1990), 789–815. (D. Ressler, M. S. Thesis, Dept. of Mech. Eng'g and Mechanics, Lehigh University, Bethlehem, PA, 1984.)

[11] Shim, G., and Wei, R. P., "Corrosion Fatigue and Electrochemical Reactions in Modified HY130 Steel," Materials Science and Engineering, 86 (1987), 121–135.

[12] Shim, G., Nakai, Y., and Wei, R. P., "Corrosion Fatigue and Electrochemical Reactions in Steels," in Basic Questions in Fatigue, ASTM STP 925, Vol. II, Am. Soc. for Testing and Materials, Philadelphia, PA (1988), 211–229.

[13] Chu, H. C., and Wei, R. P., "Stress Corrosion Cracking of High-Strength Steels in Aqueous Environments," Corrosion, 46, 6 (June 1990), 468–476; Chu, H. C., "Stress Corrosion Cracking of High-Strength Steels in Aqueous Environments," Dissertation, Lehigh University (1987).

[14] Wei, R. P., and Chiou, S., "Corrosion Fatigue Crack Growth and Electrochemical Reactions for a X-70 Linepipe Steel in Carbonate-Bicarbonate Solution," Engr. Fract. Mech., 41, 4 (1992), 463–473.

[15] Wei, R. P., "Environmentally Assisted Fatigue Crack Growth," in Advances in Fatigue Science and Technology, Kluwer Academic Publishers, Norwell, MA (1989), 221–252.

10 Science-Based Probability Modeling and Life Cycle Engineering and Management

10.1 Introduction

Material aging, through the evolution and distribution of damage (*e.g.*, by localized corrosion and corrosion fatigue), is one of the principal causes for the reduction in the reliability and margin of safety of engineered systems. It can contribute significantly to the cost of maintenance and operation and, thereby, the overall life cycle cost. To quantify materials aging and to facilitate the overall optimization of the performance, reliability, and life cycle costs of these systems (*i.e.*, for life cycle engineering and management (LCEM)) new modeling approaches are needed. Traditional (and current) approaches to engineering design are no longer adequate, because these approaches are based largely on the use of experientially based statistical methodologies and accelerated testing over periods that are well short of those of the intended service. The models developed from them are essentially parametric representations of statistical fits to the experimental data, and are effective only over the range of the underlying data. They capture, at best, the influences of the limited number of controlled (*external*) variables used in testing. Furthermore, variability associated with measurement errors (which cannot be separated from the experimental data) are incorporated into the statistical analyses, and can lead to overestimations of the uncertainty bounds. As such, simple application of known statistical techniques cannot provide the necessary tools for LCEM of engineered systems, and a different approach needs to be adopted. Here, a science (mechanistically)-based, probability-modeling approach that has been used successfully over the past decade [1–7] is presented to illustrate the modeling process and its efficacy. The overall framework and approach are described. Its use and efficacy are illustrated through two examples: first, on modeling of pitting corrosion and fatigue crack growth in aluminum alloys and its application to aging aircraft, and second, in considering the fatigue (S-N) response of a bearing steel into the very high cycle domain (*i.e.*, up to 10^{10} cycles).

10.2 Framework

Materials aging is considered in the context of its influence on the assessments of reliability, safety, availability, and maintenance of engineered systems. The framework for these assessments is depicted in Fig. 10.1. Within it, the materials-aging process is reflected specifically in the evolution and distribution of *damage* that compromise *functionality, reliability*, and *safety*. The key issues, therefore, pertain to the assessment of such a system under given sets of projected operating conditions (*i.e.*, in terms of forcing functions and environmental conditions) in relation to its *current state* or its *initial state* (either new, or after major maintenance service) and its *future state*. Such assessments are typically made through the use of a set of analysis tools, in conjunction with a comprehensive suite of diagnostic or nondestructive evaluation (NDE) tools that provide information on the current state (sizes and distribution) of damage in the system.

Assurance of reliability and continued safety, and availability, requires a quantitative assessment of the system in its '*projected future state*.' For this assessment, appropriate quantitative models are needed for estimating the accumulation of damage (in size and distribution) over its projected period of operation. The outcome of this assessment then serves as the basis for decisions on its suitability for *continued service* as reflected in Fig. 10.1 by the labels *Reliable, Conditioned Reliability*, and *Not Reliable*. A system judged to be reliable would be accepted for unrestricted operation until the next scheduled maintenance, the one with conditioned reliability would be subjected to operational constraints, and the one deemed to be

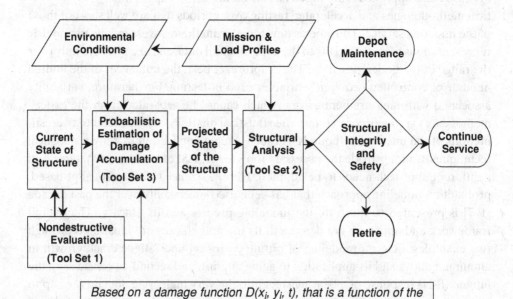

Based on a damage function $D(x_i, y_i, t)$, that is a function of the key internal (x_i) and external (y_i) variables

Figure 10.1. A simplified flow diagram for life prediction, reliability assessment and management of engineered systems [6]

unreliable would be sent for overhaul or be retired. The process labeled as *Probabilistic Estimation of Damage Accumulation* in Fig. 10.1 is the key element of this process. The confidence that can be placed on this assessment depends importantly on the robustness of the underlying models for damage evolution and distribution within a component or system. It requires the development of methods that are *predictive* and that can provide accurate estimates of the evolution and probabilistic distribution in damage over time that can be used for reliability and safety assessments and service life prediction.

10.3 Science-Based Probability Approach

10.3.1 Methodology

The requisite methodology must provide the following capabilities: (i) projection beyond typical underlying data, (ii) analyses for critical variable response, (iii) investigation into the reliability and availability of components and systems, and (iv) life cycle engineering and management of systems. Science (mechanistically)-based probability modeling, *vis-à-vis* experientially based statistical modeling, provides the structure to meet this need. A comparative assessment of these two approaches is given in [7]. The essence of science-based probability modeling of damage evolution and distribution is the formulation of a time-dependent damage function $D(x_i, y_i, t)$ that captures its functional dependence on all of the key *internal* (x_i; e.g., materials) and *external* (y_i; e.g., loading) variables, and their variability. As such, this damage function $D(x_i, y_i, t)$ accounts for its mechanistic and statistical dependence on the key random variables. It is, thereby, the foundation for time-dependent probability analyses for estimating the distribution of damage, or the distribution in service lives, that are essential for system design and management.

The development of $D(x_i, y_i, t)$ is based on scientific understanding and modeling of the mechanisms of damage nucleation and growth. The essential process for model development is shown schematically in Fig. 10.2, and is iterative. It involves the identification and confirmation of a set of key *external* and *internal* variables, and the formulation of an appropriate mechanistic (deterministic) model for $D(x_i, y_i, t)$ that express its functional dependence on these variables. The next step is to determine the probability distribution for each of the key variables in terms of either the probability density function (*pdf*), or the cumulative distribution function (*cdf*). From these functions, say the *pdfs*, a joint probability density function (*jpdf*) is constructed. The *jpdf* is then integrated with the mechanistic model to yield a science-based probability (stochastic) model. In practice, however, the stochastic results are to be derived through simulation; *e.g.*, through the use of Monte Carlo methods. The experientially based statistical methods, on the other hand, bypass the identification and quantification of the role of *internal* variables, and model development is by-and-large limited to establishment of empirical fitting functions to the experimental data.

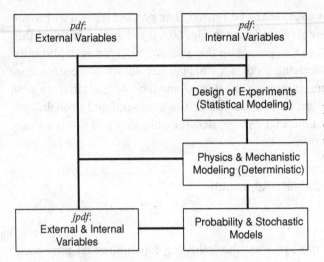

Figure 10.2. Simplified flow diagram for the development of mechanistically based probability models.

10.3.2 Comparison of Approaches

The philosophical and practical differences between the two approaches to modeling are given by Harlow and Wei [7], by using a tensile ligament instability model for creep-controlled crack growth [see 7] and by statistical least-squares fit to the experimental data. The mechanistic model is based on the recognition that crack growth is governed by the "tensile instability" (or necking failure) of ligaments in the crack-tip process zone ahead of the crack tip. These ligaments are identified with the regions of material isolated by the growth of voids nucleated at nonmetallic inclusions in high-strength steels [see 7]. In this model, the steady-state creep crack growth rate $(da/dt)_{sm}$ is related to the steady-state creep rate in the tensile ligament within the process zone and the crack growth life through the Hart-Li model for creep [see 7]. The statistical model, on the other hand, is a simple two-parameter exponential equation that fits the data in semilogarithmic space. The comparison is shown in Fig. 10.3. Note that, because the data were obtained from a very small sample of

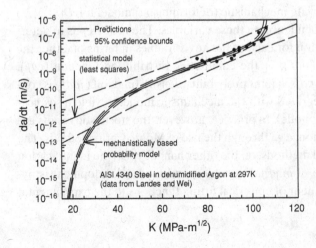

Figure 10.3. Comparison between mechanistically based probability and statistically based models for crack growth kinetics [7].

material, most of the uncertainty in the statistical model reflected errors in crack length measurements, rather than "true" variability in material properties. Note also the significant difference between the two models in the low-growth-rate region, which is of paramount importance for life prediction.

The difference in approach is self-evident. In the mechanistically based model, key *internal* and *external* variables are identified. Their variabilities are readily incorporated into the model to assess the overall variability in response. The contribution of each of the random variables on the variability in response may be readily assessed. Given the explicit functional dependence, when duly validated, it can be used to predict response beyond the range of the experimental data. The experientially based statistical model, on the other hand, represents a statistical fit to the data in which the key *internal* variables could not be identified. As such, it is incapable of capturing the functional dependence on these variables, and its usefulness is limited to the range of the experimental data. Because experimental (including measurement) errors are lumped into estimates of the fitting parameters and their variability, the quality of the subsequent reliability analyses may be overly conservative, or uncertain. A more detailed discussion of these approaches may be found in [7].

It is worth noting that a crucial difference exists in the role of experimentation between the *science-based probability* and the *experientially based statistical* approaches. For the science-based probability approach, experimentation is one of discovery and hypothesis testing to guide model formulation. For the experientially based statistical approach, on the other hand, the goal is to establish the best parametric fit to the experimental data in terms of a limited set of identifiable *external* variables. In the first case, variability arises naturally out of the randomness in the key internal and external variables, whereas the other simply captures the scatter in experimental data. In the following sections, modeling of pitting corrosion and corrosion fatigue of aluminum alloys is used to illustrate the process, and to demonstrate the efficacy and utility of the approach for estimating the evolution and distribution of damage for LCEM of engineered systems. The applicability of this approach in understanding the dichotomy between S-N and fracture mechanics approaches to corrosion fatigue is discussed. The use of this approach to understand S-N response of a high-strength-bearing steel in the very high cycle regime (up to 10^{10} cycles) is discussed.

10.4 Corrosion and Corrosion Fatigue in Aluminum Alloys, and Applications

10.4.1 Particle-Induced Pitting in an Aluminum Alloy

A simplified model for pit growth was first proposed by Harlow and Wei [2] and was used successfully to account for damage evolution in airframe aluminum alloys. For simplicity, the model assumed the pit to be hemispherical in shape, with radius *a*, and its growth (driven by an external constant-current source) would be at a

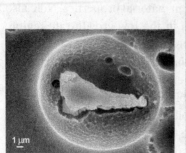
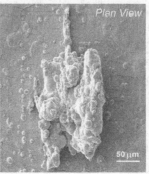

Figure 10.4. Scanning electron micrographs of (a) a particle-induced corrosion pit, and (b) the epoxy replica of a severe corrosion pit in plan (bottom) and side (elevation) view relative to the original pit in a 2024-T3 aluminum alloy sheet [9].

constant volumetric rate, obeying Faraday's law. Specifically, the pit volume V, is represented by one-half of a sphere, with $V = (2/3)\pi a^3$. The rate of pit growth, the time evolution of pit size and the time required to reach a given pit size are as follows:

$$\left.\begin{aligned}
\frac{da}{dt} &= \frac{da}{dV}\frac{dV}{dt} = \frac{1}{2\pi a^2}\frac{dV}{dt} = \frac{MI_p}{2\pi n\rho F}\frac{1}{a^2} \\
a &= \left[\frac{3MI_p}{2\pi n\rho F}t + a_o^3\right]^{1/3} \\
t &= \frac{2\pi n\rho F}{3MI_p}\left(a^3 - a_o^3\right)
\end{aligned}\right\} \tag{10.1}$$

In Eqn. (10.1), M is the molecular weight; I_p is the pitting current; n is the valency; ρ is the density; F is Faraday's constant (9.65×10^7 C/kg-mol), and a_o is the initial pit size, or the size of the initiating particle or particle cluster. For aluminum, $M = 27$ kg/kg-mol; $n = 3$; and $\rho = 2.7 \times 10^3$ kg/m^3. For particle-induced pitting, the pitting current is determined by the cathodic current density that can be supported by the particle (or cluster of particles) and its surface area.

Based on studies of pitting corrosion in 2024-T3 aluminum alloy (see Fig. 10.4, for example), it is recognized that pitting resulted naturally from dissolution of the aluminum matrix through its galvanic coupling to the constituent particles [8]. Based on this recognition, a simple, science-based model was proposed [9]. The model envisioned that a pit would be nucleated at a surface particle, in a 'contiguous cluster' of constituent particles, by galvanic corrosion of the matrix. Its continued growth would be sustained by galvanic current from other particles in the cluster that are progressively exposed at the surface of the growing pit [10–12].

For modeling, the particles are again approximated by spheres of different radii. The rate of pit growth around the surface particle of radius a_o (regime 1), and the time evolution of pit size are identical to those given in Eqn. (10.1). The pitting current I_p, however, is explicitly taken to be the product of the limiting cathodic current density i_{co} that can be supported by the particle and the surface area of the particle A_{po}, i.e., $I_p = i_{co} A_{po} = i_{co}(2\pi a_o^2)$. The particle area is taken to be one-half of the surface area of a sphere to account approximately for the increase in exposed surface as the pit grows. The extent of this initial stage of growth depends on the point at which pitting separates the particle from the alloy matrix and the time when sufficient subsurface particles are exposed to sustain continued pit growth. This "transition size" is taken as a_{tr}, and was estimated to be about three times a_o. The initial stage of growth, therefore, is explicitly given (for $a_o \leq a \leq a_{tr}$) in terms of Eqn. (10.1) as follows:

$$\left.\begin{aligned} \frac{da}{dt} &= \frac{MI_p}{2\pi n\rho F}\frac{1}{a^2} = \frac{Mi_{co}(2\pi a_o^2)}{2\pi n\rho F}\frac{1}{a^2} = \frac{Mi_{co}a_o^2}{n\rho F}\frac{1}{a^2} \\ a &= \left[\frac{3Mi_{co}a_o^2}{n\rho F}t + a_o^3\right]^{1/3} \\ t &= \frac{n\rho F}{3Mi_{co}a_o^2}\left(a^3 - a_o^3\right) \end{aligned}\right\} \quad ; \quad a_o \leq a \leq a_{tr} \qquad (10.2)$$

Following transition, the pit would continue to grow through the subsurface cluster (regime 2), with the growth supported by galvanic coupling current between the matrix (pit surface) and the exposed constituent particles at the pit surface. Because the particles vary widely in size, and composition, and electrochemical character, 'average' values are used in the model. Assuming that the constituent particles within the cluster, with an average radius $\bar{a}_p$ (micrometers), are uniformly distributed with an average density $\bar{d}_p$ (particles per millimeter square), the average number of particles $\bar{n}_p$ that are exposed on the surface of hemispherical pit of radius a (micrometers) at time t (hour) would be given by $\bar{n}_p = \bar{d}_p(2\pi a^2)$. The area of the particles that would be exposed to the electrolyte within a growing pit at time t is taken, on average, to be equal to $\bar{n}_p(2\pi\bar{a}_p^2)$. The pitting current is then:

$$I_p = i_{co}\bar{n}_p\left(2\pi\bar{a}_p^2\right) = i_{co}(\bar{d}_p \cdot 2\pi a^2)\left(2\pi\bar{a}_p^2\right) \qquad (10.3)$$

Note that the limiting cathodic current density i_{co} depends on the composition of the individual constituent particles and electrochemical conditions within the pit, both of which can also change over time. For simplicity, it was taken to be an ensemble average over the particles and was assumed to be constant over time.

The pit growth rate in regime 2 was obtained by substituting Eqn. (10.3) into the first of Eqn. (10.1), and is as follows:

$$\frac{da}{dt} = \frac{MI_p}{2\pi n\rho F}\frac{1}{a^2} = \frac{M}{n\rho F}\frac{1}{2\pi a^2}i_{co}(\bar{d}_p \cdot 2\pi a^2)\left(2\pi\bar{a}_p^2\right) = \frac{Mi_{co}\bar{d}_p}{n\rho F}\left(2\pi\bar{a}_p^2\right) \qquad (10.4)$$

Equation (10.4) indicates that the pit growth rate in this regime would be, on average, constant. The time evolution of a particle-induced corrosion pit is as follows:

$$\left.\begin{aligned} a &= a_{tr} + \frac{Mi_{co}\bar{d}_p}{n\rho F}\left(2\pi\bar{a}_p^2\right)(t - t_{tr}) \\ t &= t_{tr} + \frac{n\rho F}{Mi_{co}\bar{d}_p}\frac{1}{\left(2\pi\bar{a}_p^2\right)}(a - a_{tr}) \end{aligned}\right\} \quad ; \quad a \geq a_{tr} \qquad (10.5)$$

In this representation, the initial growth involves an increment of only about $2a_o$, and the subsequent growth through the particle cluster is at an essentially constant rate. From an engineering perspective, it is appealing to simplify it by assuming that the pit had started instead from a cluster of particles that is somewhat larger than the "gate keeper" particle. As such, the growth rate would be wholly constant. Such a model is physically plausible, and would still capture the essence of the model. The growth rate and time evolution relations are essentially those given by Eqns. (10.4) and (10.5), except that the transition pit size a_{tr} is replaced by the initial pit (cluster) size a_o.

A comparison of the models and experimental data is shown in Fig. 10.5. Pit depth measurements were made for pits that were formed in 2024-T3 aluminum alloy sheet specimens after immersion in 0.5 M NaCl solution ($[O_2] = 7$ p.p.m.) for 16 to 384 h [10]. The measured pit depths are shown as a function of exposure time compared with the model predictions in Fig. 10.5. For these comparisons, an average particle radius of 5 μm was used. A deterministic (constant) value of i_{co} of 200 μA/cm^2 was used throughout to estimate the "average" influences of particle composition, solution acidification, dealloying, and copper deposition [10–12]. For illustrative purposes, variability is considered here only through the choices in particle densities of 3,000, 1,330, and 500 particles/mm^2, with corresponding starting particle sizes of 15, 10, and 5 μm for a_{11}, a_{12}, and a_{13} and starting cluster size of 18, 14, and 10 μm for a_{21}, a_{22}, and a_{23}, respectively. It is seen that these models are in good agreement with the trend of the measured data (a_{exp}). In reality, however, variability would reflect the combined influences of variations in a_o, a_p, n_p, and i_{co}, or in appropriate combinations of these variables.

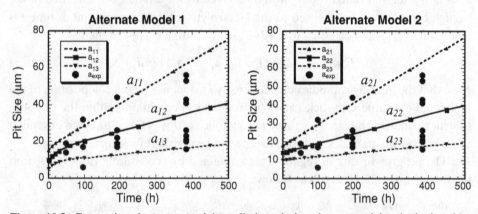

Figure 10.5. Comparison between model predictions (a_{ij}) and measured (a_{exp}) pit sizes in a 2024-T3 aluminum alloy exposed to 0.5 M NaCl solution at room temperature [10].

In these models, the pit growth rate depends on the particle radius (a_o or a_p) and density (d_p) and the limiting cathodic current density (i_{co}); these are the *internal* random variables. The cathodic current density would depend on the solution chemistry, particle composition, and temperature, all of which need to be quantified. The particle radius can range from about 1 to 30 μm, and meaningful particle densities are related to the particle size and would cover interparticle distances of two to four particle radii. The limiting cathodic current density depends on the composition of the particle and electrochemical conditions within the pit and can range from about 40 to 600 $\mu A/cm^2$. The range of variations in these parameters (a_o, a_p, n_p, and i_{co}) provides reasonable coverage of the observed variability in growth rates.

10.4.2 Impact of Corrosion and Fatigue Crack Growth on Fatigue Lives (S-N Response)

In this example, the foregoing pitting model is combined with a fatigue crack growth model to examine the contribution of each key *internal* random variable on the variability in fatigue lives, and to highlight the intimate connection between S-N and crack growth. Here, Eqn. (10.5) is used for pit growth and the following empirical equation is used for fatigue crack growth:

$$\left(\frac{da}{dN}\right) = C_c(\Delta K - \Delta K_{th})^{n_c}; \quad \Delta K = \beta \Delta\sigma a^{1/2} \tag{10.6}$$

In Eqn. (10.6), a is the crack size; N is the number of loading cycles; (da/dN) is the rate of fatigue crack growth per loading cycle; C_c is the material- and environment-dependent growth rate coefficient; and ΔK is the driving force for crack growth, given by the stress intensity factor range from linear fracture mechanics; ΔK_{th} is the fatigue crack growth threshold; n_c is the power-law exponent; β is a geometric parameter; and $\Delta\sigma$ is the (tensile) stress range. In recognition of the fact that Eqn. (10.6) is not a mechanistically based rate equation, n_c is taken to be deterministic to reflect its expected constancy in such a model and for dimensional considerations. In addition to the random variables chosen for pitting, C_c and ΔK_{th} are taken to be the *internal* random variables for crack growth. Transition for pitting to fatigue crack growth is expected to occur when the effective ΔK for the pit exceeds ΔK_{th}, and when the time-based rate of fatigue crack growth exceeds the rate of pit growth; namely,

$$(\Delta K)_{\text{pit}} \geq \Delta K_{th} \quad \text{and} \quad (da/dt)_{\text{crack}} \geq (da/dt)_{\text{pit}} \tag{10.7}$$

The time-based fatigue crack growth rate is simply $f(da/dN)$, where f is the frequency of cyclic loading.

Experimental data suggest that, in practice, transition from pitting to fatigue crack growth is determined by the second criterion in Eqn. (10.7). From Eqns. (10.4) and (10.7), the transition crack size a_{tr} may be determined by solving the following equality; namely,

$$fC_c\left(\beta\Delta\sigma a_{tr}^{1/2} - \Delta K_{th}\right)^{n_c} = \frac{Mi_{co}\overline{d}_p}{n\rho F}\left(2\pi\overline{a}_p^2\right) \tag{10.8}$$

Table 10.1. *Key random parameters and the associated Weibull cumulative distribution function parameters*

Random variable	α	β	γ	μ	cv
Initial pit radius, a_o (μm)	1.29	11.78	5.7	16.6	78%
Pitting current density, I (A/m^2)	2.6	0.56	0.5	1.0	41%
Coefficient, C_F (m/cyc)/(MPa$\sqrt{m}$)$^{3.55}$	15	9.9E-12	3.0E-11	3.95E-11	8%
Threshold driving force, ΔK_{th} (MPa$\sqrt{m}$)	2.1	0.34	0.2	0.5	50%

The number of fatigue cycles associated with pit growth (N_{pit}) and for fatigue crack growth (N_{feg}), and the overall fatigue life (N_F) in a smooth specimen are as follows:

$$N_{\text{pit}} = f t_{\text{pit}} = f \frac{n\rho F}{M i_{co} \bar{d}_p} \frac{1}{(2\pi \bar{a}_p^2)}(a_{tr} - a_o) \tag{10.9a}$$

$$N_{fcg} \approx \frac{2}{(n_c - 2)C_F \beta^2 \Delta\sigma^2 (\beta \Delta\sigma a_{tr}^{1/2} - \Delta K_{th})^{(n_c - 2)}}$$

$$\times \left[1 + \frac{(n_c - 2)\Delta K_{th}}{(n_c - 1)(\beta \Delta\sigma a_{tr}^{1/2} - \Delta K_{th})}\right]; \quad n_c > 2 \tag{10.9b}$$

$$N_F \approx N_{\text{pit}} + N_{fcg} \tag{10.9c}$$

Representing each of the *internal* random variables by a Weibull distribution, and using reasonable estimates for these values (see Table 10.1), the fatigue life sensitivity to each of the variables was determined through Monte Carlo simulation, and is shown in Fig. 10.6. Their collective impact on the distributions in fatigue lives at various stress levels is shown in Fig. 10.7. Without corrosion, the corresponding fatigue lives would have been up to three orders of magnitude longer, depending on the applied stress $\Delta\sigma$; see Eqn. (10.8). This example illustrates the importance of a mechanistically based probability approach in identifying the key random variables, and in assessing their influences on service life, and structural integrity and reliability.

Figure 10.6. Single simulation showing the sensitivity of fatigue lives to variability in each of the *internal* random variables (see Table 10.1 at $\Delta\sigma$ = 200 MPa).

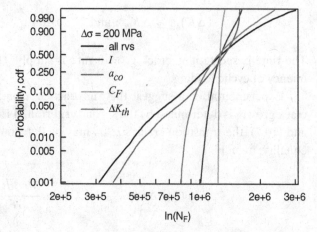

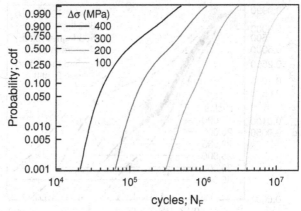

Figure 10.7. Variability in fatigue lives attributed to the *internal* random variables (see Table 10.1) at different stress levels.

10.4.3 S-N versus Fracture Mechanics (FM) Approaches to Corrosion Fatigue and Resolution of a Dichotomy

From the preceding analyses, it is clear that S-N response is significantly affected by pitting, which principally serves to truncate the early stage of fatigue crack growth and shorten fatigue life. In other words, conventional corrosion fatigue response reflects the foreshortening of corrosion-fatigue crack growth life by pitting corrosion. Because electrochemical variables strongly influence pit growth, these variables would also affect the conventional S-N data. Crack growth, on the other hand, occurs by hydrogen embrittlement and depends on the crack-tip environment, which is shielded by and large, from changes in external electrochemical variables. As such, it would be essentially independent of these variables. From this perspective, therefore, the perceived dichotomy (*i.e.*, the inconsistency in electrochemical response) between the conventional and fracture mechanics approaches to corrosion fatigue (and stress corrosion cracking) is resolved. Although the discussion here is focused on the influence of pitting corrosion on corrosion fatigue, it may be generalized to include other forms of localized corrosion, as well as stress corrosion cracking. In light of this resolution, it would be reasonable and worthwhile to re-examine the wealth of research data on corrosion fatigue over the past decades to broaden the understanding of corrosion fatigue.

10.4.4 Evolution and Distribution of Damage in Aging Aircraft

From an engineering perspective, to demonstrate the efficacy and utility of this modeling approach, a comparison was made between the model predictions and damage measured on a transport aircraft that had been in commercial service for about twenty-four years. Instead of predictions of corrosion and corrosion fatigue lives, the models were exercised through 'Monte Carlo' simulation to determine the evolution and distribution in damage size as a function of time. The results are shown in Fig. 10.8. The specifics of the analyses and comparisons are detailed elsewhere [14]. The essence of the finding is that, by using short-term laboratory data, the model

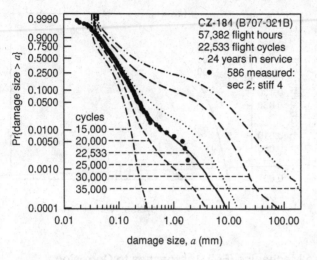

Figure 10.8. Estimated evolution and distribution of damage versus observations in the CZ-184 aircraft [14].

was able to capture the essence of the size and distribution in damage for an aircraft that had been in service for about twenty-four years (*i.e.*, for well over two orders of magnitude extrapolation in time). Through this process, the model may be used also for estimating the evolution and spatial distribution in damage over time (Fig. 10.9), either over different locations in a given structure or component, or for a single location in a group of structures or components.

10.5 S-N Response for Very-High-Cycle Fatigue (VHCF)

Considerable interest was developed in the late 1990s by the observation of unexpected S-N fatigue response at lives in the 10^8 to 10^{10} cycle range; see the papers in

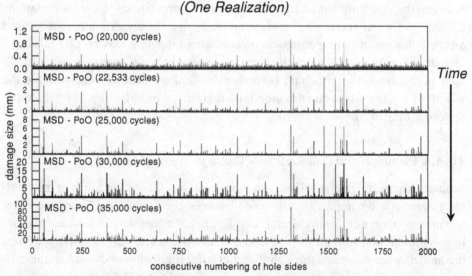

Figure 10.9. Successive simulation showing the evolution and distribution of corrosion and fatigue damage, and the formation of significant areas of multi-site damage (MSD) over 1,000 fastener holes for the CZ-184 aircraft [14].

15.0kV x200 SEI   ⊢——⊣   50μm 15.0kV x2400 SEI   ⊢—⊣   2μm

Figure 10.10. Scanning electron micrographs of crack nucleation at a typical inclusion, high and low magnifications, respectively [15].

the special session on gigacycle fatigue in [15]. The response is reflected by a lower "endurance limit" relative to that observed by the conventional procedure of testing to only 10^7 to 10^8 cycles, and by the prominence of subsurface crack nucleation at internal inclusion particles in the high-cycle domain. These internal nucleation sites have been dubbed "fish eyes" because of their darkened appearance in optical microscopy (see Fig. 10.10). The precise mechanisms for this high-cycle response are not fully understood. In [16], Murakami summarizes the view that attributes the behavior to the local concentration of dissolved hydrogen at the crack-nucleating inclusions. Other possible contributors include the influences of residual stresses (*surface* versus *interior*), environment (*external* versus *internal*), or both.

Based on the studies summarized above (see Section 10.4), it is reasonable to assume that the conventional S-N response for steels can also be related directly to the crack growth life (*i.e.*, the number of fatigue cycles required to grow a crack from its nucleus to failure). This approach was applied to assess computationally the influences on S-N response by "surface" residual stress and its surface-to-interior distribution or the effects of environment, as well as the probabilistic influences of the variability in the size of crack nuclei and in other material properties. In [17], a crack growth-based probability description for fatigue life prediction into the gigacycle range that explicitly incorporates the effects from internal and external damage is proposed. A connection between the S-N and crack growth behaviors was established and demonstrated. Through this description, the S-N response and the associated variability in fatigue lives are linked to key random variables that are explicitly identified in the crack growth model; namely, the initial surface damage sizes, the initial internal damage (inclusion) sizes, the fatigue crack growth rate (or power-law) coefficient, and the fatigue crack threshold ΔK (ΔK_{th}). The identification and quantification of these random variables are vital for probabilistic estimation and prediction of fatigue life. The model is assessed through comparisons with an extensive set of fatigue life data for SUJ2 steel [17] (Figs. 10.11 and 10.12).

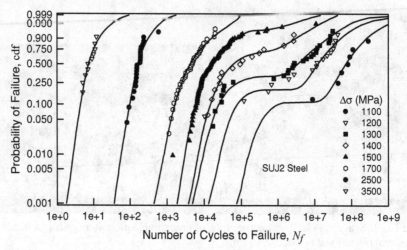

Figure 10.11. S-N data for SUJ2 steel along with *cdfs* computed from a fatigue crack growth model [15].

The analyses were focused on the role of internal residual stresses, with an assumed probability distribution that served to favor crack nucleation at the specimen surface at the higher stresses, and internal nucleation near the endurance limit. The results (Fig. 10.11) show the agreement between the estimated and observed distributions in fatigue lives at different stress levels [17]. Note that the distributions at the lower stress levels (1100 to 1400 MPa) could not have been established through conventional statistical procedures. It is recognized that SUJ2 is a high-strength-bearing steel that would be susceptible to environmentally enhanced crack growth in moist environments. As such, crack nucleation and growth from the surface sites would be enhanced relative to those from the interior sites, particularly when the stress intensity, or K, level is above that for "stress corrosion cracking" (or K_{Iscc}). This environmental influence would naturally account for the preferred

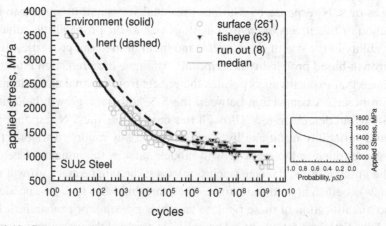

Figure 10.12. Schematic of the median characteristics for an alternative interpretation of the very-high-cycle fatigue S-N behavior [15].

nucleation of fatigue crack growth from the surface sites at the higher stresses. For the size range of the crack nuclei, the appropriate stress level is near the endurance limit of this steel. As such, the observed behavior may be attributed to this environmental influence, which is illustrated in Fig. 10.12. The precise causes for the observed response are to be identified through well designed critical experiments.

10.6 Summary

In this chapter, the need for science-based probability modeling of damage evolution and distribution for use in LCEM of modern, high-value-added engineered systems was highlighted. The approach and its efficacy were illustrated and demonstrated through selected examples. These findings served to highlight plausible causes for the observed responses, identify the potential key random variables, and provide guidance for further investigations. To support this transformation in approach, fundamental changes in the basic and applied science and engineering must be made. Experimentation must be focused on discovery, hypothesis testing, and validation to support the identification and quantification of key random variables and model development, *vis-à-vis*, on phenomenology *per se*.

REFERENCES

[1] Harlow, D. G., and Wei, R. P., "A Mechanistically Based Approach to Probability Modeling for Corrosion Fatigue Crack Growth," Engr. Frac. Mech., 45, 1 (1993), 79–88.

[2] Harlow, D. G., and Wei, R. P., "Probability Approach for Corrosion and Corrosion Fatigue Life," J. of the Am. Inst. of Aeronautics and Astronautics, 32, 10 (1994), 2073–2079.

[3] Wei, R. P., Masser, D., Liu, H., and Harlow, D. G., "Probabilistic Considerations of Creep Crack Growth," Mater. Sci. & Engr., A189 (1994), 69–76.

[4] Harlow, D. G., Lu, H.-M., Hittinger, J. A., Delph, T. J., and Wei, R. P., "A Three-Dimensional Model for the Probabilistic Intergranular Failure of Polycrystalline Arrays," Modelling Simul. Mater. Sci. Eng., 4 (1996), 261–279.

[5] Harlow, D. G., and Wei, R. P., "A Probability Model for the Growth of Corrosion Pits in Aluminum Alloys Induced by Constituent Particles," Engr. Frac. Mech., 59, 3 (1998), 305–325.

[6] Harlow, D. G., and Wei, R. P., "Probabilities of Occurrence and Detection of Damage in Airframe Materials," Fat. & Fract. of Engr. Matls & Structures, 22 (1999), 427–436.

[7] Harlow, D. G., and Wei, R. P., "A Critical Comparison between Mechanistically Based Probability and Statistically Based Modeling for Materials Aging," Mater. Sci. & Eng. (2002), 278–284.

[8] Liao, C.-M., Chen, G. S., and Wei, R. P., "A Technique for Studying the 3-Dimensional Shape of Corrosion Pits," Scripta Mater., 35, 11 (1996), 1341–1346.

[9] Chen, G. S., Wan, K.-C., Gao, M., Wei, R. P., and Flournoy, T. H., "Transition From Pitting to Fatigue Crack Growth – Modeling of Corrosion Fatigue Crack Nucleation in a 2024-T3 Aluminum Alloy," Matls Sci. and Engr., A219 (1996), 126–132.

[10] Dolley, E. J., Lee, B., and Wei, R. P., "The Effect of Pitting Corrosion on Fatigue Life," Fat. & Fract. of Engr. Mat. & Structures, 23 (2000), 555–560.

[11] Gao, M., Feng, C. R., and Wei, R. P., "An AEM Study of Constituent Particles in Commercial 7075-T6 and 2024-T3 Alloys," Metall. Mater. Trans., 29A (1998), 1145–1151.

[12] Wei, R. P., Liao, C.-M., and Gao, M., "A Transmission Electron Microscopy Study of Constituent Particle-Induced Corrosion in 7075-T6 and 2024-T3 Aluminum Alloys," Metall. Mater. Trans., 29A (1998), 1153–1160.

[13] Wei, R. P., "Corrosion/Corrosion Fatigue and Life-Cycle Management," Mat. Sci. Research International, 7, 3 (2001), 147–156.

[14] Wei, R. P., and Harlow, D. G., "Corrosion-Enhanced Fatigue and Multiple-Site Damage," AIAA Journal, 41, 10 (2003), 2045–2050.

[15] Special Session: Giga-Cycle Fatigue, in A. F. Blom, ed., Fatigue 2002, 5, Engineering Materials Advisory Services Ltd., West Midlands, UK (2002), 2927–2994.

[16] Murakami, Y., Mechanism of Fatigue Failure in Ultra-long Life Regime and Application to Fatigue Design, in A. F. Blom, ed., Fatigue 2002, 5, Engineering Materials Advisory Services Ltd., West Midlands, UK, (2002), 2927–2938.

[17] Harlow, D. G., Wei, R. P., Sakai, T., and Oguma, N., "Crack Growth Based Probability Modeling of S-N Response for High-Strength Steel," Inter. J. of Fatigue, 28 (2006), 1479–1485.

APPENDIX

Publications by R. P. Wei and Colleagues

OVERVIEW/GENERAL

Wei, R. P., "Application of Fracture Mechanics to Stress Corrosion Cracking Studies," in *Fundamental Aspects of Stress Corrosion Cracking*, NACE, Houston, TX (1969), 104.

Wei, R. P., and Speidel, M. O., "Phenomenological Aspects of Corrosion Fatigue, Critical Introduction," *Corrosion Fatigue: Chemistry, Mechanics and Microstructure*, NACE-2 (1972), 279.

Wei, R. P., Novak, S. R., and Williams, D. P., "Some Important Considerations in the Development of Stress Corrosion Cracking Test Methods," Advisory Group for Aerospace Research and Development (AGARD) Conf. Proc. No. 98, *Specialists Meeting on Stress Corrosion Testing Methods* (1971), and Materials Research and Standards, ASTM, 12, 9 (1972), 25.

McEvily, A. J., and Wei, R. P., "Fracture Mechanics and Corrosion Fatigue," *Corrosion Fatigue: Chemistry, Mechanics and Microstructure*, NACE-2 (1972), 281.

Wei, R. P., and Speidel, M. O., "Phenomenological Aspects of Corrosion Fatigue, Critical Introduction," *Corrosion Fatigue: Chemistry, Mechanics and Microstructure*, NACE-2 (1972), 279.

Wei, R. P., "The Effect of Temperature and Environment on Subcritical Crack Growth," *Fracture Prevention and Control*, ASM Materials/Metalworking Technology Series No. 3 (1974), 73.

Wei, R. P., "Contribution of Fracture Mechanics to Subcritical Crack Growth Studies," in *Linear Fracture Mechanics*, G. C. Sih, R. P. Wei, and F. Erdogan, eds., ENVO Publishing Co., Lehigh Valley, PA (1976), 287–302.

Wei, R. P., "Environmental Considerations in Fatigue and Fracture of Constructional Steels," in *New Horizons in Construction Materials*, Vol. I, H.-Y. Fang, ed., ENVO Publishing Co., Lehigh Valley, PA (1977).

Wei, R. P., "On Understanding Environment Enhanced Fatigue Crack Growth-A Fundamental Approach," in *Fatigue Mechanisms*, ASTM STP 675, J. T. Fong, ed., American Society for Testing & Materials, Philadelphia, PA (1979), 816–840.

Wei, R. P., "Fatigue Crack Growth in Aqueous and Gaseous Environments," in *Environmental Degradation of Engineering Materials in Aggressive Environments*, Vol. 2, M. R. Louthan, Jr., R. P. McNitt, and R. D. Sisson, Jr., eds., Virginia Polytechnic Institute, Blacksburg, VA (1981), 73–81.

Wei, R. P., and Novak, S. R., "Interlaboratory Evaluation of K_{Iscc} Measurement Procedures for Steels: A Summary," in *Environment Sensitive Fracture: Evaluation and Comparison of Test Methods*, ASTM Special Technical Publication (STP) 821, S. W. Dean, E. N. Pugh, and G. M. Ugiansky, eds., American Society for Testing and Materials, Philadelphia, PA (1984), 75–79.

Wei, R. P., "Chemical and Microstructural Aspects of Corrosion Fatigue Crack Growth," in *FRACTURE Mechanics: Microstructure and Micromechanisms*, Proceedings of ASM 1987 Materials Science Seminar, S. V. Nair, J. K. Tien, R. C. Bates, and O. Buck, eds., ASM International, Metals Park, OH (1989), 229–254.

Wei, R. P., "Environmentally Assisted Fatigue Crack Growth," in *Advances in Fatigue Science and Technology*, M. Branco and L. Guerra Rosa, eds., Kluwer Academic Publishers, Norwell, MA (1989), 221–252.

Wei, R. P., "Electrochemical Considerations of Crack Growth in Ferrous Alloys," *Advances in Fracture Research*, Proceedings of Seventh International Conference on Fracture, Houston, TX, March (1989), K. Salama, K. Ravi-Chandar, D. M. R. Taplin, and P. Rama Rao, eds., Permagon Press, Oxford, UK (1989), 1525–1544.

Wei, R. P., and Harlow, D. G., "Materials Considerations in Service Life Prediction," Proceedings of DOE Workshop on Aging of Energy Production and Distribution Systems, Rice University, Houston, TX, October 11–12 (1992), M. M. Carroll and P. D. Spanos, eds., Appl. Mech. Rev., 46, 5 (1993), 190–193.

Wei, R. P., "Corrosion Fatigue: Science and Engineering," in *Recent Advances in Corrosion Fatigue*, Sheffield, UK April 16–17, 1997.

Wei, R. P., "Progress in Understanding Corrosion Fatigue Crack Growth," in *High Cycle Fatigue of Structural Materials*, W. O. Soboyejo and T. S. Srivatsan, eds., The Minerals, Metals and Materials Society, Warrendale, PA (1997), 79–80.

Wei, R. P., "Aging of Airframe Aluminum Alloys: From Pitting to Cracking," Proceedings of Workshop on Intelligent NDE Sciences for Aging and Futuristic Aircraft, FAST Center for Structural Integrity of Aerospace Systems, The University of Texas at El Paso, El Paso, TX, September 30–October 2, 1997, C. Ferregut, R. Osegueda, and A. Nuñez, eds. (1997), 113–122.

Wei, R. P., "A Perspective on Environmentally Assisted Crack Growth in Steels," Proceedings of International Conference on Environmental Degradation of Engineering Materials, Gdansk-Jurata, Poland, September 19–23 (1999).

FRACTURE

Baker, A. J., Lauta, F. J., and Wei, R. P., "Relationships Between Microstructure and Toughness in Quenched and Tempered Ultrahigh-Strength Steels," ASTM STP 370 (1965), 3.

Wei, R. P., "Fracture Toughness Testing in Alloy Development," ASTM STP 381 (1965), 279.

Wei, R. P., and Lauta, F. J., "Measuring Plane-Strain Fracture Toughness with Carbonitrided Single-Edge-Notch Specimens," Materials Research and Standards, ASTM, 5, 6 (1965), 305.

Birkle, A. J., Wei, R. P., and Pellissier, G. E., "Analysis of Plane-Strain Fracture in a Series of 0.45C-Ni-Cr-Mo Steels with Different Sulfur Contents," Trans. ASM, 59, 4 (1966), 981.

STRESS CORROSION CRACKING/HYDROGEN-ENHANCED CRACK GROWTH

Wei, R. P., "Application of Fracture Mechanics to Stress Corrosion Cracking Studies," in *Fundamental Aspects of Stress Corrosion Cracking*, NACE (1969), 104.

Wei, R. P., Novak, S. R., and Williams, D. P., "Some Important Considerations in the Development of Stress Corrosion Cracking Test Methods," AGARD Conf. Proc. No. 98, *Specialists Meeting on Stress Corrosion Testing Methods* (1971), and Materials Research and Standards, ASTM, 12, 9 (1972), 25.

Wei, R. P., Klier, K., Simmons, G. W., and Chornet, E., "Hydrogen Adsorption and Diffusion, and Subcritical-Crack Growth in High–Strength Steels and Nickel Base Alloys," First Annual Report, NASA Grant NGR 39-007-067, January (1973).

Gangloff, R. P., and Wei, R. P., "Gaseous Hydrogen Assisted Crack Growth in 18 Nickel Maraging Steels," Scripta Metallurgica, 8 (1974), 661.

Wei, R. P., Klier, K., Simmons, G. W., Gangloff, R. P., Chornet, E., and Kellerman, R., "Hydrogen Adsorption and Diffusion, and Subcritical-Crack Growth in High-Strength Steels and Nickel-Base Alloys," Lehigh University Report IFSM-74-63, Final Report to NASA Lewis Research Center for Grant NGR 39-007-067 (June 1974).

Chou, Y. T., and Wei, R. P., "Elastic Interactions of a Moving Crack with Vacancies and Solute Atoms," Acta Metallurgical, 23 (1975), 279.

Hudak, S. J., and Wei, R. P., "Hydrogen Enhanced Crack Growth in 18 Ni Maraging Steels," Metallurgical Transactions A, 7A, (1976), 235–241.

Wei, R. P., and Simmons, G. W., "A Technique for Determining the Elemental Composition of Fracture Surfaces Produced by Crack Growth in Hydrogen and in Water Vapor," Scripta Metallurgica, 10, 2 (1976), 153–157.

Chou, Y. T., Tsao, K. Y., and Wei, R. P., "On the Elastic Interaction of a Broberg Crack with Vacancies and Solute Atoms," Materials Science and Engineering, 24 (1976), 101–107.

Pao, P. S., and Wei, R. P., "Hydrogen Assisted Crack Growth in 18Ni(300) Maraging Steel," Scripta Metallurgica, 11 (1977), 515–520.

Gangloff, R. P., and Wei, R. P., "Gaseous Hydrogen Embrittlement of High Strength Steels," Metallurgical Transactions A, 8A (1977), 1043–1053.

Dwyer, D. J., Simmons, G. W., and Wei, R. P., "A Study of the Initial Reaction of Water Vapor with Fe(001) Surface," Surface Sci., 64 (1977), 617–632.

Simmons, G. W., and Wei, R. P., "Environment Enhanced Fatigue Crack Growth in High-Strength Steels," in Stress Corrosion Cracking and Hydrogen Embrittlement of Iron Based Alloys, J. Hochmann, J. Slater, and R. W. Staehle, eds., NACE, Houston, TX (1978), 751–765.

Chou, Y. T., Wu, R. S., and Wei, R. P., "Time-Dependent Flow of Solute Atoms Near a Crack Tip," Scripta Metallurgica, 12 (1978), 249–254.

Ganglolff, R. P., and Wei, R. P., "Fractographic Analysis of Gaseous Hydrogen Induced Cracking in 18Ni Maraging Steel," Fractography in Failure Analysis, ASTM STP 645 (1978), 87–106.

Chan, N. H., Klier, K., and Wei, R. P., "A Preliminary Investigation of Hart's Model in Hydrogen Embrittlement in Maraging Steels," Scripta Metallurgica, 12 (1978), 1043–1046.

Simmons, G. W., Pao, P. S., and Wei, R. P., "Fracture Mechanics and Surface Chemistry Studies of Subcritical Crack Growth in AISI 4340 Steel," Metallurgical Transactions A, 9A (1978), 1147–1158.

Williams, III, D. P., Pao, P. S., and Wei, R. P., "The Combined Influence of Chemical, Metallurgical and Mechanical Factors on Environment Assisted Cracking," in Environment Sensitive Fracture of Engineering Materials, Z. A. Foroulis, ed., The Minerals, Metals, and Masterials Society-American Institute of Mining, Metallurgical, and Petroleum Engineers (TMS-AIME) (1979), 3–15.

Lu, M., Pao, P. S., Chan, N. H., Klier, K., and Wei, R. P., "Hydrogen Assisted Crack Growth in AISI 4340 Steel," Proceedings Japan Institute and Metals International Symposium-2, Hydrogen in Metals (1980), 449–452.

Chan, N. H., Klier, K., and Wei, R. P., "Hydrogen Isotope Exchange Reactions Over the AISI 4340 Steel," Proceedings JIMIS-2, Hydrogen in Metals (1980), 305–308.

Wei, R. P., "Rate Controlling Processes and Crack Growth Response," in Hydrogen Effects in Metals, I. M. Bernstein and Anthony W. Thompson, eds., The Metallurgical Society of AIME, Warrendale, PA (1981), 677–690.

Lu, M., Pao, P. S., Weir, T. W., Simmons, G. W., and Wei, R. P., "Rate Controlling Processes for Crack Growth in Hydrogen Sulfide for an AISI 4340 Steel," Metallurgica Transactions A, 12A (1981), 805–811.

Hudak, Jr., S. J., and Wei, R. P., "Consideration of Nonsteady-State Crack Growth in Materials Evaluation and Design," Int'l. J. Pres. & Piping, 9 (1981), 63–74.

Wei, R. P., Klier, K., Simmons, G. W., and Chou, Y. T., "Fracture Mechanics and Surface Chemistry Investigations of Environment-Assisted Crack Growth," in *Hydrogen Embrittlement and Stress Corrosion Cracking*, Ronald Gibala, et al., eds., American Society for Metals, Metals Park, OH (1984), 103.

Gao, M., Lu, M., and Wei, R. P., "Crack Paths and Hydrogen-Assisted Crack Growth Response in AISI 4340 Steel," Metallurgical Transactions A, 15A, (April 1984), 735–746.

Wei, R. P., Gao, M., and Pao, P. S., "The Role of Magnesium in CF and SCC of 7000 Series Aluminum Alloys," Scripta Metallurgica, 18, 11 (1984), 1195–1198.

Wei, R. P., and Novak, S. R., "Interlaboratory Evaluation of K_{Iscc} Measurement Procedures for Steels: A Summary," in *Environment Sensitive Fracture: Evaluation and Comparison of Test Methods*, ASTM STP 821, S. W. Dean, E. N. Pugh, and G. M. Ugiansky, eds., American Society for Testing and Materials, Philadelphia, PA (1984), 75–79.

Gao, M., and Wei, R. P., "Quasi-Cleavage and Martensite Habit Plane," Acta Metallurgica, 32, 11 (1984), 2115–2124.

Wei, R. P., and Gao, M., "Chemistry, Microstructure and Crack Growth Response," in *Hydrogen Degradation of Ferrous Alloys*, R. A. Oriani, J. P. Hirth, and S. Smialowski, eds., Noyes Publications, Park Ridge, NJ (1985), 579–603.

Wei, R. P., "Synergism of Mechanics, Mechanisms and Microstructure in Environmentally Assisted Crack Growth," in *FRACTURE: Interactions of Microstructure, Mechanisms and Mechanics*, J. M. Wells and J. D. Landes, eds., The Metallurgical Society of AIME, Warrendale, PA (1985), 75–88.

Gao, M., and Wei, R. P., "A "Hydrogen Partitioning" Model for Hydrogen Assisted Crack Growth," Metallurgical Transactions A, 16A (1985), 2039–2050.

Gangloff, R. P., and Wei, R. P., "Small Crack-Environment Interactions: The Hydrogen Embrittlement Perspective," in *Small Fatigue Cracks*, R. O. Ritchie and J. Lankford, eds., The Metallurgical Society of AIME, Warrendale, PA (1986), 239–263.

Wei, R. P., and Simmons, G. W., "Modeling of Environmentally Assisted Crack Growth," in *Environment Sensitive Fracture of Metals and Alloys*, R. P. Wei, D. J. Duquette, T. W. Crooker, and A. J. Sedriks, eds., Office of Naval Research, Arlington, VA (1987), 63–77.

Wei, R. P., Gao, M., and Xu, P. Y., "Peak Bare-Surface Densities Overestimated in Straining and Scratching Electrode Experiments," J. Electrochem. Soc., 136, 6 (1989), 1835–1836.

Chu, H. C., and Wei, R. P., "Stress Corrosion Cracking of High-Strength Steels in Aqueous Environments," Corrosion, 46, 6 (1990), 468–476.

Wei, R. P., and Gao, M., "Hydrogen Embrittlement and Environmentally Assisted Crack Growth," in *Hydrogen Effects on Material Behavior*, N. R. Moody and A. W. Thompson, eds., The Minerals, Metals & Materials Society, Warrendale, PA (1990), 789–816.

Gao, M., Boodey, J. B., and Wei, R. P., "Hydrides in Thermally Charged Alpha-2 Titanium Aluminides," Scripta Met. et Matl., 24 (1990), 2135–2138.

Wei, R. P., and Gao, M., "Further Observations on the Validity of Bare Surface Current Densities Determined by the Scratched Electrode Technique," J. Electrochem. Soc., 138, 9 (1991), 2601–2606.

Gao, M., Boodey, J. B., and Wei, R. P., "Misfit Strains and Mechanism for the Precipitation of Hydrides in Thermally Charged Alpha-2 Titanium Aluminides," in *Environmental Effects on Advanced Materials*, R. H. Jones and R. E. Ricker, eds., The Minerals, Metals and Materials Society, Warrendale, PA (1991), 47–55.

Wei, R. P., and Alavi, A., "In Situ Fracture Techniques for Studying Transient Reactions With Bare Steel Surfaces," J. of the Electrochem. Soc., 138, 10 (1991), 2907–2912.

Boodey, J. B., Gao, M., and Wei, R. P., "Hydrogen Solubility and Hydride Formation in a Thermally Charged Gamma-Based Titanium Aluminide," in *Environmental Effects on Advanced Materials*, R. H. Jones and R. E. Ricker, eds., The Minerals, Metals and Materials Society, Warrendale, PA (1991), 57–65.

Wei, R. P., and Gao, M., "Distribution of Initial Current Between Bare and Filmed Surfaces (What is Being Measured in a Scratched Electrode Test?)," Corrosion, 47, 12 (1992), 948–951.

Gao, M., Boodey, J. B., Wei, R. P., and Wei, W., "Hydrogen Solubility and Microstructure of Hastelloy X," Scripta Met. et Mater., 26 (1992), 63–68.

Gao, M., Boodey, J. B., Wei, R. P., and Wei, W., "Hydrogen Solubility and Microstructure of Gamma Based Titanium Aluminides," Scripta Met. et Mater., 27 (1992), 1419–1424.

Chen, S., Gao, M., and Wei, R. P., "Phase Transformation and Cracking During Aging of an Electrolytically Charged Fe18Cr12Ni Alloy at Room Temperature," Scripta Met. et Mater., 28 (1993), 471–476.

Valerio, P., Gao, M., and Wei, R. P., "Environmental Enhancement of Creep Crack Growth in Inconel 718 by Oxygen and Water Vapor," Scripta Metall. et Mater., 30, 10 (1994), 1269–1274.

Gao, M., Dunfee, W., Wei, R. P., and Wei, W., "Thermal Fatigue of Gamma Titanium Aluminide in Hydrogen," in *Fatigue and Fracture of Ordered Intermetallic Materials: I*, W. O. Soboyejo, T. S. Srivatsan, and D. L. Davidson, eds., The Minerals, Metals & Materials Society, Warrendale, PA (1994), 225–237.

DEFORMATIOM (CREEP) CONTROLLED CRACK GROWTH

Li, C. Y., Talda, P. M., and Wei, R. P., unpublished research, Applied Research Laboratory, U. S. Steel Corp., Monroeville, PA (1966).

Landes, J. D., and Wei, R. P., "Kinetics of Subcritical Crack Growth and Deformation in a High Strength Steel," J. Eng'g Materials and Technology, ASME, Ser. H, 95 (1973), 1–9.

Landes, J. D., and Wei, R. P., "The Kinetics of Subcritical Crack Growth under Sustained Loading," Int'l. J. of Fracture, 9 (1973), 277–286.

Yin, H., Gao, M., and Wei, R. P., "Deformation and Subcritical Crack Growth under Static Loading." Matl's Sci. & Eng'g., A119 (1989), 51–58.

Wei, R. P., Masser, D., Liu, H. W., and Harlow, D. G., "Probabilistic Considerations of Creep Crack Growth," Materials Science and Engineering, A189 (1994), 69–76.

OXYGEN-ENHANCED CRACK GROWTH

Gao, M., and Wei, R. P., "Precipitation of Intragranular $M_{23}C_6$ Carbides in a Nickel Alloy: Morphology and Crystallographic Feature," Scripta Met. et Mater., 30, 8 (1994), 1009–1014.

Pang, X. J., Dwyer, D. J., Gao, M., Valerio, P., and Wei, R. P., "Surface Enrichment and Grain Boundary Segregation of Niobium in Inconel 718 Single-and Poly-Crystals," Scripta Metall. et Materialia, 31, 3 (1994), 345–350.

Valerio, P., Gao, M., and Wei, R. P., "Environmental Enhancement of Creep Crack Growth in Inconel 718 by Oxygen and Water Vapor," Scripta Metall. et Mater., 30, 10 (1994), 1269–1274.

Dwyer, D. J., Pang, X. J., Gao, M., and Wei, R. P., "Surface Enrichment of Niobium on Inconel 718 (100) Single Crystals," Applied Surf. Sci., 81 (1994), 229–235.

Gao, M., and Wei, R. P., "Grain Boundary γ'' Precipitation and Niobium Segregation in Inconel 718," Scripta Metall. et Mater, 32, 7 (1995), 987–990.

Gao, M., Dwyer, D. J., and Wei, R. P., "Niobium Enrichment and Environmental Enhancement of Creep Crack Growth in Nickel-Base Superalloys," Scripta Metall. et Mater., 32, 8 (1995), 1169–1174.

Liu, H., Gao, M., Harlow, D. G., and Wei, R. P., "Grain Boundary Character, and Carbide Size and Spatial Distribution in a Ternary Nickel Alloy," Scripta Metall. et Mater. 32, 11 (1995), 1807–1812.

Gao, M., Dwyer, D. J., and Wei, R. P., "Chemical and Microstructural Aspects of Creep Crack Growth in Inconel 718 Alloy," in *Superalloys 718, 625, 706 and Various Deivatives*, E. A. Loria, ed., The Minerals, Metals & Materials Society, Warrendale, PA (1995), 581–592.

Lu, H.-M., Delph, T. J., Dwyer, D. J., Gao, M., and Wei, R. P., "Environmentally-Enhanced Cavity Growth in Nickel and Nickel-Based Alloys," Acta Mater., 44, 8 (1996), 3259–3266.

Gao, M., Chen, S., and Wei, R. P., "Preferential Coarsening of γ'' Precipitates in Inconel 718 During Creep," Metall. Mater. Trans., 27A (1996), 3391–3398.

Gao, M., Chen, S. F., Chen, G. S., and Wei, R. P., "Environmentally Enhanced Crack Growth in Nickel-Based Alloys at Elevated Temperatures," in Elevated Temperature Effects on Fatigue and Fracture, ASTM STP 1297, R. S. Piascik, R. P. Gangloff, and A. Saxena, eds., American Society for Testing and Materials, West Conshohocken, PA (1997), 74–84.

Chen, G. S., Aimone, P. R., Gao, M., Miller, C. D., and Wei, R. P., "Growth of Nickel-Base Superalloy Bicrystals by the Seeding Technique with Modified Bridgman Method," J. of Crystal Growth, 179 (1997), 635–646.

Gao, M., and Wei, R. P., "Grain Boundary Niobium Carbides in Inconel 718," Scripta Mater., 37, 12 (1997), 1843–1849.

Wei, R. P., Liu, H., and Gao, M., "Crystallographic Features and Growth of Creep Cavities in a Ni-18Cr-18Fe Alloy," Acta Mater., 46, 1 (1998), 313–325.

Chen, S.-F., and Wei, R. P., "Environmentally Assisted Crack Growth in a Ni-18Cr-18Fe Ternary Alloy at Elevated Temperatures," Matls Sci. & Engr., A256 (1998), 197–207.

Wei, R. P., Liu, H., and Gao, M., "Crystallographic Features and Growth of Creep Cavities in a Ni-18Cr-18Fe Alloy," Acta Mater., 46, 1 (1998), 313–325.

Chen, S.-F., and Wei, R. P., "Environmentally Assisted Crack Growth in a Ni-18Cr-18Fe Ternary Alloy at Elevated Temperatures," Matls Sci. & Engr., A256 (1998), 197–207.

Rong, Y., Chen, S., Hu, G., Gao, M., and Wei, R. P., "Prediction and Characterization of Variant Electron Diffraction Patterns for γ'' and δ Precipitates in INCONEL 718 Alloy," Met. & Mater. Trans., 30A (1999), 2297–2303.

Wei, R. P., Liu, H., and Gao, M., "Crystallographic Features and Growth of Creep Cavities in a Ni-18Cr-18Fe Alloy," Acta Mater., 46, 1 (1998), 313–325.

Chen, S.-F., and Wei, R. P., "Environmentally Assisted Crack Growth in a Ni-18Cr-18Fe Ternary Alloy at Elevated Temperatures," Matls Sci. & Engr., A256 (1998), 197–207.

Iwashita, C. H., and Wei, R. P., "Coarsening of Grain Boundary Carbides in a Nickel-Base Ternary Alloy During Creep," Acta Mater., 48 (2000), 3145–3156.

Miller, C. F., Simmons, G. W., and Wei, R. P., "High Temperature Oxidation of Nb, NbC and Ni_3Nb and Oxygen Enhanced Crack Growth," Scripta Mater., 42 (2000), 227–232.

Wei, R. P., "Oxygen Enhanced Crack Growth in Nickel-based P/M Superalloys," Proceedings of Symposium on Advanced Technologies for Superalloy Affordability, TMS 2000 Annual Meeting, Nashville, TN, 12–16 March (2000).

Wei, R. P., and Huang, Z., "Influence of Dwell Time on Fatigue Crack Growth in Nickel-Based Superalloys," Mat. Sci. and Eng., A336 (2002), 209–214.

Miller, C. F., Simmons, G. W., and Wei, R. P., "Mechanism for Oxygen Enhanced Crack Growth in Inconel 718," Scripta Mater., 44 (2001), 2405–2410.

Huang, Z., Iwashita, C., Chou, I., and Wei, R. P., "Environmentally Assisted, Sustained-Load Crack Growth in Powder Metallurgy Nickel-Based Superalloys," Metallurgical and Materials Trans A, 33A (2002), 1681–1687.

Miller, C. F., Simmons, G. W., and Wei, R. P., "Evidence for Internal Oxidation During Oxygen Enhanced Crack Growth in P/M Ni-based Superalloys," Scripta Materialia 48 (2003), 103–108.

Wei, R. P., Miller, C., Huang, Z., Simmons, G. W., and Harlow, D. G., "Oxygen Enhanced Crack Growth in Nickel-based Super Alloys and Materials Damage Prognosis," Engineering Fracture Mechanics, 76, 5 (2009), 715–727.

FATIGUE/CORROSION FATIGUE

Wei, R. P., and Baker, A. J., "Observation of Dislocation Loop Arrays in Fatigued Polycrystalline Pure Iron," Phil. Mag., 11, 113, (1965), 1087.

Wei, R. P., and Baker, A. J., "A Metallographic Study of Iron Fatigue in Cyclic Strain at Room Temperature," Phil. Mag., 12, 119 (1965), 1005.

Li, C.-Y., Talda, P. M., and Wei, R. P., "The Effect of Environments on Fatigue–Crack Propagation in an Ultra-High-Strength Steel," Int'l. J. Fract. Mech., 3 (1967), 29.

Wei, R. P., Talda, P. M., and Li, C.-Y., "Fatigue-Crack Propagation in Some Ultra-High-Strength Steels," ASTM STP 415 (1967), 460.

Spitzig, W. A., and Wei, R. P., "A Fractographic Investigation of the Effect of Environment on Fatigue-Crack Propagation in an Ultrahigh-Strength Steel," Trans. ASM, 60 (1967), 279.

Spitzig, W. A., Talda, P. M., and Wei, R. P., "Fatigue-Crack Propagation and Fractographic Analysis of 18Ni(250) Maraging Steel Tested in Argon and Hydrogen Environments," Eng'g. Fract. Mech., 1 (1968), 155.

Wei, R. P., "Fatigue-Crack Propagation in a High-Strength Aluminum Alloy," Int'l. J. Fract. Mech., 4, 2 (1968), 159.

Wei, R. P., and Landes, J. D., "The Effect of D_2O on Fatigue-Crack Propagation in a High-Strength Aluminum Alloy," Int'l. J. Fract. Mech., 5 (1969), 69.

Wei, R. P., and Landes, J. D., "Correlation Between Sustained-Load and Fatigue Crack Growth in High Strength Steels," Materials Research and Standards, ASTM 9, 7 (1969), 25.

Wei, R. P., "Some Aspects of Environment-Enhanced Fatigue-Crack Growth," Eng'g. Fract. Mech., 1, 4 (1970), 633.

Spitzig, W. A., and Wei, R. P., "Fatigue-Crack Propagation in Modified 300-Grade Maraging Steel," Eng'g. Fract. Mech., 1, 4 (1970), 719.

Feeney, J. A., McMillan, J. C., and Wei, R. P., "Environmental Fatigue Crack Propagation of Aluminum Alloys at Low Stress Intensity Levels," Metallurgical Transactions, 1 (1970), 1741.

Jonas, O., and Wei, R. P., "An Exploratory Study of Delay in Fatigue-Crack Growth," Int'l. J. Fract. Mech., 7 (1971), 116.

Ritter, D. L., and Wei, R. P., "Fractographic Observations of Ti-6Al-4V Alloy Fatigued in Vacuum," Metallurgical Transactions, 2 (1971), 3229.

Wei, R. P., and Ritter, D. L., "The Influence of Temperature on Fatigue Crack Growth in a Mill Annealed Ti-6Al-4V Alloy," J. Materials, ASTM, 7, 2 (1972), 240.

Gallagher, J. P., and Wei, R. P., "Corrosion Fatigue Crack Propagation Behavior in Steels," Corrosion Fatigue: Chemistry, Mechanics and Microstructure, NACE-2 (1972), 409.

Miller, G. A., Hudak, S. J., and Wei, R. P., "The Influence of Loading Variables on Environment-Enhanced Fatigue Crack Growth in High Strength Steels," J. of Testing and Evaluation, ASTM, 1 (1973), 524.

Wei, R. P., and Shih, T. T., "Delay in Fatigue Crack Growth," Int't. J. Fract. Mech., 10, 1 (1974), 77; also as Wei, R. P., Shih, T. T., and Fitzgerald, J. H., "Load Interaction Effects on Fatigue Crack Growth in Ti-6Al-4V Alloy," NASA CR-2239 (April 1973).

Shih, T. T., and Wei, R. P., "A Study of Crack Closure in Fatigue," J. Eng'g. Fract. Mech., 6 (1974), 19; also as Shih, T. T., and Wei, R. P., "A Study of Crack Closure in Fatigue," NASA CR-2319 (October 1973).

Fitzgerald, J. H., and Wei, R. P., "A Test Procedure for Determining the Influence of Stress Ratio on Fatigue Crack Growth," J. Testing and Evaluation, ASTM, 2, 2 (1974), 67.

Shih, T. T., and Wei, R. P., "Load and Environment Interactions in Fatigue Crack Growth," Proceedings – International Conference on Prospects of Fracture Mechanics, Delft, Netherlands (1974), 231.

Shih, T. T., and Wei, R. P., "Effect of Specimen Thickness on Delay in Fatigue Crack Growth," J. of Testing and Evaluation, ASTM, 3, 1 (1975), 46.

Shih, T. T., and Wei, R. P., "Influences of Chemical and Thermal Environments on Delay in a Ti-6Al-4V Alloy," in Fatigue Crack Growth Under Spectrum Loads, ASTM STP 595, American Soc. of Testing and Materials, Philadelphia, PA (1976), 113–124.

Unangst, K. D., Shih, T. T., and Wei, R. P., "Crack Closure in 2219-T851 Aluminum Alloy," Eng'g. Fract. Mech., 9 (1977), 725–734.

Wei, R. P., "Fracture Mechanics Approach to Fatigue Analysis in Design," J. Eng'g. Mat'l. & Tech., 100 (1978), 113–120.

Simmons, G. W., Pao, P. S., and Wei, R. P., "Fracture Mechanics and Surface Chemistry Studies of Subcritical Crack Growth in AISI 4340 Steel," Metallurgical Transactions A, 9A (1978), 1147–1158.

Pao, P. S., Wei, W., and Wei, R. P., "Effect of Frequency on Fatigue Crack Growth Response of AISI 4340 Steel in Water Vapor," *Environment Sensitive Fracture of Engineering Materials*, TMS-AIME, Z. A. Foroulis, ed. (1979), 565–580.

Williams, III, D. P., Pao, P. S., and Wei, R. P., "The Combined Influence of Chemical, Metallurgical and Mechanical Factors on Environment Assisted Cracking," in *Environment Sensitive Fracture of Engineering Materials*, TMS-AIME, Z. A. Foroulis, ed. (1979), 3–15.

Wei, R. P., "On Understanding Environment Enhanced Fatigue Crack Growth – A Fundamental Approach," in *Fatigue Mechanisms*, ASTM STP 675, J. T. Fong, ed., American Society for Testing & Materials, Philadelphia, PA (1979), 816–840.

Wei, R. P., Wei, W., and Miller, G. A., "Effect of Measurement Precision and Data-Processing Procedures on Variability in Fatigue-Crack Growth Rate Data," J. of Testing & Evaluation, JTEVA, 7, 2 (1979), 90–95.

Brazill, R. L., Simmons, G. W., and Wei, R. P., "Fatigue Crack Growth in 2-1/4Cr-1Mo Steel Exposed to Hydrogen Containing Gases," J. Eng'g. Mat'l. & Tech., Trans. ASME, 101 (1979), 199–204.

Wei, R. P., Pao, P. S., Hart, R. G., Weir, T. W., and Simmons, G. W., "Fracture Mechanics and Surface Chemistry Studies of Fatigue Crack Growth in an Aluminum Alloy," Metallurgical Transactions A, 11A (1980), 151–158.

Weir, T. W., Simmons, G. W., Hart, R. G., and Wei, R. P., "A Model for Surface Reaction and Transport Controlled Fatigue Crack Growth," Scripta Metallurgica, 14 (1980), 357–364.

Wei, R. P., Fenelli, N. E., Unangst, K. D., and Shih, T. T., "Fatigue Crack Growth Response Following a High-Load Excursion in 2219-T851 Aluminum Alloy," J. Eng'g. Mat'l. & Tech., Trans. of ASME, 102, 3 (1980), 280–292.

Wei, R. P., and Simmons, G. W., "Recent Progress in Understanding Environment Assisted Fatigue Crack Growth," Int'l. J. of Fract., 17, 2 (1981), 235–247.

Wei, R. P., "Rate Controlling Processes and Crack Growth Response," in *Hydrogen Effects in Metals*, I. M. Bernstein and A. W. Thompson, eds., The Metallurgical Society of AIME, Warrendale, PA (1981), 677–690.

Lu, M., Pao, P. S., Weir, T. W., Simmons, G. W., and Wei, R. P., "Rate Controlling Processes for Crack Growth in Hydrogen Sulfide for an AISI 4340 Steel," Metallurgica Transactions A, 12A (1981), 805–811.

Wei, R. P., "Fatigue Crack Growth in Aqueous and Gaseous Environments," in *Environmental Degradation of Engineering Materials in Aggressive Environments*, Vol. 2, M. R. Louthan, Jr., R. P. McNitt, and R. D. Sisson, Jr., eds., Virginia Polytechnic Institute, Blacksburg, VA (1981), 73–81.

Wei, R. P., and Simmons, G. W., "Surface Reactions and Fatigue Crack Growth," in *FATIGUE: Environment and Temperature Effects*, J. J. Burke and V. Weiss, eds., Sagamore Army Materials Research Conference Proceedings, 27 (1983), 59–70.

Shih, T.-H., and Wei, R. P., "The Effects of Load Ratio on Environmentally Assisted Fatigue Crack Growth," Eng'g. Fract. Mech., 18, 4 (1983), 827–837.

Wei, R. P., and Gao, M., "Reconsideration of the Superposition Model For Environmentally Assisted Fatigue Crack Growth," Scripta Metallurgica, 17 (1983), 959–962.

FATIGUE MECHANISMS

Advances in Quantitative Measurement of Fatigue Damage, ASTM STP 811, J. Lankford, D. L. Davidson, W. L. Morris, and R. P. Wei, eds., American Society for Testing and Materials, Philadelphia, PA (1983).

Wei, R. P., and Shim, G., "Fracture Mechanics and Corrosion Fatigue," in *Corrosion Fatigue*, ASTM STP 801, T. W. Crooker and B. N. Leis, eds., American Society for Testing and Materials, Philadelphia, PA (1983), 5–25.

Gao, S. J., Simmons, G. W., and Wei, R. P., "Fatigue Crack Growth and Surface Reactions For Titanium Alloys Exposed to Water Vapor," Mat'ls. Sci. & Eng'g., 62 (1984), 65–78.

Wei, R. P., "Electrochemical Reactions and Fatigue Crack Growth Response," in *Corrosion in Power Generating Equipment*, M. O. Speidel and A. Atrens, eds., Plenum Press, NY (1984), 169–174.

Wei, R. P., Shim G., and Tanaka, K., "Corrosion Fatigue and Modeling," in *Embrittlement by the Localized Crack Equipment*, R. P. Gangloff, ed., The Metallurgical Society of AIME, Warrendale, PA (1984), 243–263.

Wei, R. P., Gao, M., and Pao, P. S., "The Role of Magnesium in CF and SCC of 7000 Series Aluminum Alloys," Scripta Metallurgica, 18, 11 (1984), 1195–1198.

Tanaka, K., and Wei, R. P., "Growth of Short Fatigue Cracks in HY-130 Steel in 3.5% NaCl Solution," Engr. Fract. Mech., 21, 2 (1985), 293–305.

Gao, M., Pao, P. S., and Wei, R. P., "Role of Micromechanisms in Corrosion Fatigue Crack Growth in a 7075-T651 Aluminum Alloy," in *Fracture: Interactions of Microstructure, Mechanisms and Mechanics*, J. M. Wells and J. D. Landes, eds., The Metallurgical Society of AIME, Warrendale, PA (1985), 303–319.

Wei, R. P., "Synergism of Mechanics, Mechanisms and Microstructure in Environmentally Assisted Crack Growth," in *Fracture: Interactions of Microstructure, Mechanisms and Mechanics*, J. M. Wells and J. D. Landes, eds., The Metallurgical Society of AIME, Warrendale, PA (1985), 75–88.

Wei, R. P., Simmons, G. W., and Pao, P. S., "Environmental Effects on Fatigue Crack Growth B. Specific Environments," in *Metals Handbook, Mechanical Testing, 8*, 9th edition, American Society for Metals, Metals Park, OH (1985), 403.

Pao, P. S., Gao, M., and Wei, R. P., "Environmentally Assisted Fatigue Crack Growth in 7075 and 7050 Aluminum Alloys," Scripta Metallurgica, 19 (1985), 265–270.

Pao, P. S., and Wei, R. P., "Hydrogen-Enhanced Fatigue Crack Growth in Ti6Al-2Sn-4Zr-2Mo-0.1Si," in *Titanium: Science and Technology*, G. Lutjering, U. Zwicker, and W. Bunk, eds., FRG: Deutsche Gesellschaft Fur Metallkunde e.V. (1985), 2503.

Nakai, Y., Tanaka, K., and Wei, R. P., "Short-Crack Growth in Corrosion Fatigue for a High Strength Steel," Eng'g. Fract. Mech., 24 (1986), 443–444.

Tanaka, K., Akiniwa, Y., Nakai, Y., and Wei, R. P., "Modeling of Small Fatigue Crack Growth Interacting With Grain Boundary," Eng'g. Fract. Mech., 24 (1986), 803–819.

Thomas, J. P., Alavi, A., and Wei, R. P., "Correlation Between Electrochemical Reactions With Bare Surfaces and Corrosion Fatigue Crack Growth in Steels," Scripta Metall., 20 (1986), 1015–1018.

Wei, R. P., "Environmental Considerations in Fatigue Crack Growth," Proceedings, International Conference on Fatigue of Engineering Materials and Structures, Sheffield, England, September 15–19 (1986), IMechE, 9, The Institution of Mechanical Engineering, London (1986), 339–346.

Gangloff, R. P., and Wei, R. P., "Small Crack-Environment Interactions: The Hydrogen Embrittlement Perspective," in *Small Fatigue Cracks*, R. O. Ritchie and J. Lankford, eds., The Metallurgical Society of AIME, Warrendale, PA (1986), 239–263.

Wei, R. P., "Corrosion Fatigue Crack Growth," in *Microstructure and Mechanical Behaviour of Materials*, Vol. II, Engineering Materials Advisory Services, Warley, UK (1986), 507–526.

Wei, R. P., and Simmons, G. W., "Modeling of Environmentally Assisted Crack Growth," in *Environment Sensitive Fracture of Metals and Alloys*, R. P. Wei, D. J. Duquette, T. W. Crooker, and A. J. Sedriks, eds., Office of Naval Research, Arlington, VA (1987), 63–77.

Shim, G., and Wei, R. P., "Corrosion Fatigue and Electrochemical Reactions in Modified HY130 Steel," Mat'l. Sci. & Eng'g., 86 (1987), 121–135.

Wei, R. P., "Electrochemical Reactions and Corrosion Fatigue Crack Growth," in *Mechanical Behavior of Materials – V*, M. G. Yan, S. H. Zhang, and Z. M. Zheng, eds., Pergamon Press, Beijing (1987), 129–140.

Wei, R. P., "Environmentally Assisted Fatigue Crack Growth," in FATIGUE '87, Vol. III, R. O. Ritchie and E. A. Starke, Jr., eds., Engineering Materials Advisory Services, Warley, UK (1987), 1541–1560.

Nakai, Y., Alavi, A., and Wei, R. P., "Effects of Frequency and Temperature on Short Fatigue Crack Growth in Aqueous Environments," Met. Trans. A, 19A (1988), 543–548.

Pao, P. S., Gao, M., and Wei, R. P., "Critical Assessment of the Model for Transport-Controlled Fatigue Crack Growth," in *Basic Questions in Fatigue*, ASTM STP 925, Vol. II, American Society for Testing and Materials, Philadelphia, PA (1988), 182–195.

Shim, G., Nakai, Y., and Wei, R. P., "Corrosion Fatigue and Electrochemical Reactions in Steels," in *Basic Questions in Fatigue*, ASTM STP 925, Vol. II, American Society for Testing and Materials, Philadelphia, PA (1988), 211–229.

Gao, M., Pao, P. S., and Wei, R. P., "Chemical and Metallurgical Aspects of Environmentally Assisted Fatigue Crack Growth in 7075-T651 Aluminum Alloy," Met. Trans. A, 19A (1988), 1739.

Wei, R. P., "Corrosion Fatigue: Science and Engineering," Japan Society of Mechanical Engineers, 91, 841 (1988), 8–13 (in Japanese).

Wei, R. P., "Corrosion Fatigue Crack Growth and Reactions With Bare Steel Surfaces," Paper 569, Proceedings of Corrosion 89, New Orleans, LA, April 17–21 (1989).

Kondo, Y., and Wei, R. P., "Approach On Quantitative Evaluation of Corrosion Fatigue Crack Initiation Condition," in *International Conference on Evaluation of Materials Performance in Severe Environments*, EVALMAT 89, Vol. 1, Kobe, Japan, November 20–23 (1989), The Iron and Steel Institute of Japan, Tokyo 100, Japan (1989), 135–142.

R. P., Wei, "Mechanistic Considerations of Corrosion Fatigue of Steels," in *International Conference on Evaluation of Materials Performance in Severe Environments*, EVALMAT 89, Vol. 1, Kobe, Japan, November 20–23 (1989), The Iron and Steel Institute of Japan, Tokyo, Japan (1989), 71–85.

Thomas, J. P., and Wei, R. P., "Corrosion Fatigue Crack Growth of Steels in Aqueous Solutions – I. Experimental Results & Modeling the Effects of Frequency and Temperature," Matls. Sci. & Engr., A159 (1992), 205–221.

Thomas, J. P., and Wei, R. P., "Corrosion Fatigue Crack Growth of Steels in Aqueous Solutions – II. Modeling the Effects of Delta K," Matls. Sci. & Engr., A159 (1992), 223–229.

Gao, M., Chen, S., and Wei, R. P., "Crack Paths, Microstructure and Fatigue Crack Growth in Annealed and Cold-Rolled AISI 304 Stainless Steels," Met. Trans. A, 23A (1992), 355–371.

Wei, R. P., and Chiou, S., "Corrosion Fatigue Crack Growth and Electrochemical Reactions for a X-70 Linepipe Steel in Carbonate-Bicarbonate Solution," Engr. Fract. Mech., 41, 4 (1992), 463–473.

Gao, M., and Wei, R. P., "Morphology of Corrosion Fatigue Cracks Produced in 3.5% NaCl Solution and in Hydrogen for a High Purity Metastable Austenitic (Fe18Cr12Ni) Steel," Scripta Met. et Mater., 26, 8 (1992), 1175–1180.

Wei, R. P., and Gao, M., "Micromechanism for Corrosion Fatigue Crack Growth in Metastable Austenitic Stainless Steels," in *Corrosion-Deformation Interactions*, T. Magnin and J. M. Gras, eds., Proc. CDI '92, Fontainebleau, France, Les Editions de Physique, Les Ulis, France (1993), 619–629.

Chen, G. S., Gao, M., Harlow, D. G., and Wei, R. P., "Corrosion and Corrosion Fatigue of Airframe Aluminum Alloys," FAA/NASA International Symposium on Advanced Structural Integrity Methods for Airframe Durability and Damage Tolerance, NASA Conference Publication 3274, Langley Research Center, Hampton, VA (1994), 157–173.

Wan, K.-C., Chen, G. S., Gao, M., and Wei, R. P., "Corrosion Fatigue of a 2024-T3 Aluminum Alloy in the Short Crack Domain," Internat. J. of Fracture, 69 (3) (1994), R63–R67.

Gao, M., Chen, S., and Wei, R. P., "Electrochemical and Microstructural Considerations of Fatigue Crack Growth in Austenitic Stainless Steels," 36th Mechanical Working and Steel

Processing Conference, Vol. XXXII, October 1994, Baltimore, MD, Iron and Steel Society, Inc., Warrendale, PA (1995), 541–549.

Harlow, D. G., Cawley, N. R., and Wei, R. P., "Spatial Statistics of Particles and Corrosion Pits in 2024-T3 Aluminum Alloy," Proceedings of Canadian Congress of Applied Mechanics, May 28–June 2 (1995), Victoria, British Columbia, 116–117.

Burynski, Jr., R. M., Chen, G.-S., and Wei, R. P., "Evolution of Pitting Corrosion in a 2024-T3 Aluminum Alloy," (1995) ASME International Mechanical Engineering Congress and Exposition on Structural Integrity in Aging Aircraft, San Francisco, CA, 47, C. I. Chang and C. T. Sun, eds., The American Society of Mechanical Engineers, New York, NY (1995), 175–183.

Chen, G. S., Gao, M., and Wei, R. P., "Microconstituent-Induced Pitting Corrosion in a 2024-T3 Aluminum Alloy," CORROSION, 52, 1 (1996), 8–15.

Chen, S., Gao, M., and Wei, R. P., "Hydride Formation and Decomposition in Electrolytically Charged Metastable Austenitic Stainless Steels," Metallurgical and Materials Transactions, 27A, 1 (1996), 29–40.

Wei, R. P., and Harlow, D. G., "Corrosion and Corrosion Fatigue of Airframe Materials," U.S. Department of Transportation, Federal Aviation Administration, DOT/FAA/AR-95/76, February (1996), Final Report, National Technical Information Service, Springfield, VA (1996).

Wei, R. P., Gao, M., and Harlow, D. G., "Corrosion and Corrosion Fatigue Aspects of Aging Aircraft," Proceedings of Air Force 4th Aging Aircraft Conference, United States Air Force Academy, CO, July 9–11 (1996).

Chen, G. S., Wan, K.-C., Gao, M., Wei, R. P., and Flournoy, T. H., "Transition From Pitting to Fatigue Crack Growth – Modeling of Corrosion Fatigue Crack Nucleation in a 2024-T3 Aluminum Alloy," Matls Sci. and Engr., A219 (1996), 126–132.

Liao, C.-M., Chen, G. S., and Wei, R. P., "A Technique for Studying the 3-Dimensional Shape of Corrosion Pits," Scripta Mater., 35, 11 (1996), 1341–1346.

Chen, G. S., Liao, C.-M., Wan, K.-C., Gao, M., and Wei, R. P., "Pitting Corrosion and Fatigue Crack Nucleation," in Effects of the Environment on the Initiation of Crack Growth, ASTM STP 1298, W. A. Van Der Sluys, R. S. Piascik, and R. Zawierucha, eds., American Society for Testing and Materials, Philadelphia, PA (1997), 18–33.

Wei, R. P., "Corrosion Fatigue: Science and Engineering," in Recent Advances in Corrosion Fatigue, Sheffield, UK, April 16–17, (1997).

Wei, R. P., "Progress in Understanding Corrosion Fatigue Crack Growth," in High Cycle Fatigue of Structural Materials, W. O. Soboyejo and T. S. Srivatsan, eds., The Minerals, Metals and Materials Society, Warrendale, PA (1997), 79–80.

Wei, R. P., "Aging of Airframe Aluminum Alloys: From Pitting to Cracking," Proceedings of Workshop on Intelligent NDE Sciences for Aging and Futuristic Aircraft, FAST Center for Structural Integrity of Aerospace Systems, The University of Texas at El Paso, El Paso, TX, C. Ferregut, R. Osegueda, and A. Nuñez, eds., September 30–October 2 (1997), 113–122.

Liao, C.-M., Olive, J. M., Gao, M., and Wei, R. P., "In Situ Monitoring of Pitting Corrosion in a 2024 Aluminum Alloy," CORROSION, 54, 6 (1998), 451–458.

Gao, M., Feng, C. R., and Wei, R. P., "An AEM Study of Constituent Particles in Commercial 7075-T6 and 2024-T3 Alloys," Metall. Mater. Trans., 29A (1998), 1145–1151.

Wei, R. P., Liao, C.-M., and Gao, M., "A Transmission Electron Microscopy Study of Constituent Particle-Induced Corrosion in 7075-T6 and 2024-T3 Aluminum Alloys," Metall. Mater. Trans., 29A (1998), 1153–1160.

Harlow, D. G., and Wei, R. P., "A Probability Model for the Growth of Corrosion Pits in Aluminum Alloys Induced by Constituent Particles," Engr. Frac. Mech., 59, 3 (1998), 305–325.

Liao, C. M., Olive, J. M., Gao, M., and Wei, R. P., "In Situ Monitoring of Pitting Corrosion in Aluminum Allog 2024," Corrosion 45, n. 6, 1998, 451–458.

Wan, K.-C., Chen, G. S., Gao, M., and Wei, R. P., "Interactions between Mechanical and Environmental Variables for Short Fatigue Cracks in a 2024-T3 Aluminum Alloy in 0.5 M NaCl Solutions," Metallurgical and Materials Transactions, Part A, 31(13), (2000), 1025–1034.

Dolley, E. J., and Wei, R. P., "Importance of Chemically Short-Crack-Growth on Fatigue Life," 2nd Joint NASA/FAA/DoD Conference on Aging Aircraft, Williamsburg, VA, 31 August–3 September 1998, NASA/CP-1999-208982/PART2, Charles E. Harris, ed. (1999), 679–687.

Liao, C.-M., and Wei, R. P., "Galvanic Coupling of Model Alloys to Aluminum – A Foundation for Understanding Particle-Induced Pitting in Aluminum Alloys," Electrochimica Acta, 45 (1999), 881–888.

Liao, C.-M., and Wei, R. P., "Pitting Corrosion Process and Mechanism of 2024-T3 Aluminum Alloys," China Steel Technical Report, No. 12 (1998), 28–40.

Wei, R. P., and Harlow, D. G., "Corrosion and Corrosion Fatigue of Aluminum Alloys – An Aging Aircraft Issue," Proceedings of The Seventh International Fatigue Conference (FATIGUE '99), June 8–12 (1999), Beijing, China.

Dolley, E. J., and Wei, R. P., "The Effect of Frequency of Chemically Short-Crack-Growth Behavior & Its Impact on Fatigue Life," Proceedings of Third Joint FAA/DoD/NASA Conference on Aging Aircraft, Albuquerque, NM, September 20–23 (1999).

Wei, R. P., "A Perspective on Environmentally Assisted Crack Growth in Steels," Proceedings of International Conference on Environmental Degradation of Engineering Materials, Gdansk-Jurata, Poland, September 19–23, (1999).

Liao, C.-M., Olive, J. M., Gao, M., and Wei, R. P., [a]In-Situ Monitoring of Pitting Corrosion in Aluminum Alloy 2024," Corrosion, 54, 6 (1998), 451–458.

Wan, K.-C., Chen, G. S., Gao, M., and Wei, R. P., "Interactions between Mechanical and Environmental Variables for Short Fatigue Cracks in a 2024-T3 Aluminum Alloy in 0.5 M NaCl Solutions," Metall. Mater. Trans. A, 31A (2000), 1025–1034.

Dolley, E. J., and Wei, R. P., "Importance of Chemically Short-Crack-Growth on Fatigue Life," 2nd Joint NASA/FAA/DoD Conference on Aging Aircraft, Williamsburg, VA, August 31–September 3, 1998, NASA/CP-1999-208982/PART2, Charles E. Harris, ed. (1999), 679–687.

Liao, C.-M., and Wei, R. P., "Pitting Corrosion Process and Mechanism of 2024-T3 Aluminum Alloys," China Steel Technical Report 12 (1998), 28–40.

Liao, C.-M., and Wei, R. P., "Galvanic Coupling of Model Alloys to Aluminum – A Foundation for Understanding Particle-Induced Pitting in Aluminum Alloys," Electrochimica Acta, 45 (1999), 881–888.

Dolley, E. J., and Wei, R. P., "The Effect of Frequency of Chemically Short-Crack-Growth Behavior & Its Impact on Fatigue Life," Proceedings of Third Joint FAA/DoD/NASA Conference on Aging Aircraft, Albuquerque, NM, September 20–23 (1999).

Dolley, E. J., Lee, B., and Wei, R. P., "The Effect of Pitting Corrosion on Fatigue Life," Fat. & Fract. of Engr. Mat. & Structures, 23 (2000), 555–560.

Wan, K.-C., Chen, G. S., Gao, M., and Wei, R. P., "Interactions between Mechanical and Environmental Variables for Short Fatigue Cracks in a 2024-T3 Aluminum Alloy in 0.5 M NaCl Solutions," Metall. Mater. Trans. A, 31A (2000), 1025–1034.

Dolley, E. J., and Wei, R. P., "Importance of Chemically Short-Crack-Growth on Fatigue Life," 2nd Joint NASA/FAA/DoD Conference on Aging Aircraft, Williamsburg, VA, August 31–September 3, 1998, NASA/CP-1999-208982/PART2, Charles E. Harris, ed. (1999), 679–687.

Wei, R. P., "A Model for Particle-Induced Pit Growth in Aluminum Alloys," Acta Mater., Elsevier Science Ltd., 44 (2001), 2647–2652.

Wei, R. P., "Corrosion and Corrosion Fatigue in Perspective," Proceedings from Chemistry and Electrochemistry of Stress Corrosion Cracking: A Symposium Honoring the Contributions of R. W. Staehle, R. H. Jones, ed., The Minerals, Metals and Materials Society, Warrendale, PA (2001).

Wei, R. P., "Environmental Considerations for Fatigue Cracking," Blackwell Science Ltd. Fatigue Fract Engng Mater Struct 24 (2002), 845–854.

Papakyriacou, M., Mayer, H., Fuchs, U., Stanzl-Tschegg, S. E., and Wei, R. P., "Influence of Atmospheric Moisture on Slow Fatigue Crack Growth at Ultrasonic Frequency in Aluminum and Magnesium Alloys," Blackwell Science Ltd. Fatigue Fract Engng Mater Struct 25 (2002), 795–804.

CERAMICS/INTERMETALLICS

Gao, M., Dunfee, W., Wei, R. P., and Wei, W., "Thermal Fatigue of Gamma Titanium Aluminide in Hydrogen," in *Fatigue and Fracture of Ordered Intermetallic Materials: I*, W. O. Soboyejo, T. S. Srivatsan, and D. L. Davidson, eds., The Minerals, Metals & Materials Society, Warrendale, PA (1994), 225–237.

Dunfee, W., Gao, M., Wei, R. P., and Wei, W., "Hydrogen Enhanced Thermal Fatigue of γ-Titanium Aluminide," Scripta Metall. et Mater., 33, 2 (1995), 245–250.

Gao, M., Dunfee, W., Wei, R., and Wei, W., "Thermal Mechanical Fatigue of Gamma Titanium Aluminide in Hydrogen and Air," in *Fatigue and Fracture of Ordered Intermetallic Materials: II*, W. O. Soboyejo, T. S. Srivatsan, and R. O. Ritchie, eds., The Minerals, Metals & Materials Society, Warrendale, PA (1995), 3–15.

Yin, H., Gao, M., and Wei, R. P., "Phase Transformation and Sustained-Load Crack Growth in $ZrO_2 + 3$ mol% Y_2O_3: Experiments and Kinetic Modeling," Acta Metall. et Mater., 43, 1 (1995), 371–382.

Gao, M., Dunfee, W., Miller, C., Wei, R. P., and Wei, W., "Thermal Fatigue Testing System for the Study of Gamma Titanium Aluminides in Gaseous Environments," in Thermal-Mechanical Fatigue Behavior of Materials, Vol. 2, ASTM STP 1263, M. J. Verrilli and M. G. Castelli, eds., American Society for Testing and Materials, West Conshohocken, PA (1996), 174–186.

Gao, M., Dunfee, W., Wei, R. P., and Wei, W., "Environmentally Enhanced Thermal-Fatigue Cracking of a Gamma-Based Titanium Aluminide Alloy," Proceedings of 124th International Symposium on Gamma Titanium Aluminides VII: Microstructure and Mechanical Behavior, Las Vegas, NV, Y.-W. Kim, et al., eds., The Minerals, Metals and Materials Society, Warrendale, PA (1995), 911–918.

Boodey, J. B., Gao, M., Wei, W., and Wei, R. P., "Hydrogen Occlusion and Hydride Formation in Titanium Aluminides," Proceedings of 124th International Symposium on Gamma Titanium Aluminides VII: Microstructure and Mechanical Behavior, Las Vegas, NV, Y.-W. Kim, et al., eds., The Minerals, Metals and Materials Society, Warrendale, PA (1995), 101–108.

Dunfee, W., Gao, M., Wei, R. P., and Wei, W., "Hydrogen Enhanced Thermal Fatigue of γ-Titanium Aluminide," Scripta Metall. et Mater., 33, 2 (1995), 245–250.

MATERIAL DAMAGE PROGNOSIS/LIFE CYCLE ENGINEERING

Harlow, D. G., and Wei, R. P., "A Mechanistically Based Approach to Probability Modeling for Corrosion Fatigue Crack Growth," Engr. Frac. Mech., 45, 1 (1993), 79–88.

Harlow, D. G., and Wei, R. P., "A Mechanistically Based Probability Approach for Predicting Corrosion and Corrosion Fatigue Life," in *ICAF Durability and Structural Integrity of Airframes*, Vol. I, A. F. Blom, ed., Engineering Meterials Advisory Services, Warley, UK (1993), 347–366.

Harlow, D. G., and Wei, R. P., "A Dominant Flaw Probability Model for Corrosion and Corrosion Fatigue," in *Corrosion Control Low-Cost Reliability*, 5B, Proceedings of the 12th International Corrosion Congress, Houston, TX (1993), 3573–3586.

Wei, R. P., and Harlow, D. G., "Materials Considerations in Service Life Prediction," Proceedings of DOE Workshop on Aging of Energy Production and Distribution Systems, Rice University, Houston, TX, October 11–12, 1992, M. M. Carroll and P. D. Spanos, eds., Appl. Mech. Rev., 46, 5 (1993), 190–193.

Harlow, D. G., and Wei, R. P., "Probability Approach for Corrosion and Corrosion Fatigue Life," J. of the Am. Inst. of Aeronautics and Astronautics, 32, 10 (1994), 2073–2079.

Wei, R. P., Masser, D., Liu, H., and Harlow, D. G., "Probabilistic Considerations of Creep Crack Growth," Mater. Sci. & Engr., A189 (1994), 69–76.

Wei, R. P., and Harlow, D. G., "A Mechanistically Based Probability Approach for Life Prediction," Proceedings of International Symposium on Plant Aging and Life Predictions of Corrodible Structures, T. Shoji and T. Shibata, eds., NACE International, Houston, TX (1997), 47–58.

Harlow, D. G., and Wei, R. P., "Probability Modelling for the Growth of Corrosion Pits," (1995) ASME International Mechanical Engineering Congress and Exposition on Structural Integrity in Aging Aircraft, San Francisco, CA, 47, C. I. Chang and C. T. Sun, eds., The American Society of Mechanical Engineers, New York, NY (1995), 185–194.

Harlow, D. G., Lu, H.-M., Hittinger, J. A., Delph, T. J., and Wei, R. P., "A Three-Dimensional Model for the Probabilistic Intergranular Failure of Polycrystalline Arrays," Modelling Simul. Mater. Sci. Eng., 4 (1996), 261–279.

Wei, R. P., "Life Prediction: A Case for Multi-Disciplinary Research," in Fatigue and Fracture Mechanics, Vol. 27, ASTM STP 1296, R. S. Piascik, J. C. Newman, and N. E. Dowling, eds., American Society for Testing and Materials, Philadelphia, PA (1997), 3–24.

Cawley, N. R., Harlow, D. G., and Wei, R. P., "Probability and Statistics Modeling of Constituent Particles and Corrosion Pits as a Basis for Multiple-Site Damage Analysis," FAA-NASA Symposium on Continued Airworthiness of Aircraft Structures, DOT/FAA/AR-97/2, II, National Technical Information Service, Springfield, VA (1997), 531–542.

Wei, R. P., Li, C., Harlow, D. G., and Flournoy, T. H., "Probability Modeling of Corrosion Fatigue Crack Growth and Pitting Corrosion," ICAF 97, Fatigue in New and Ageing Aircraft, Edinburgh, Scotland, Vol. I, R. Cook and P. Poole, eds., Engineering Materials Advisory Services, Warley, UK (1997), 197–214.

Harlow, D. G., and Wei, R. P., "Probabilistic Aspects of Aging Airframe Materials: Damage versus Detection," Proceedings of the Third Pacific Rim International Conference on *Advanced Materials and Processes* (PRICM 3), M. A. Imam, R. DeNale, S. Hanada, Z. Zhong, and D. N. Lee, eds., Honolulu, Hawaii, July 12–16, 1998, The Minerals, Metals & Materials Society, Warrendale, PA (1998), 2657–2666.

Harlow, D. G., and Wei, R. P., "Aging of Airframe Materials: Probability of Occurrence Versus Probability of Detection," 2nd Joint NASA/FAA/DoD Conference on Aging Aircraft, Williamsburg, VA, August 31–September 3 1998, NASA/CP-1999-208982/PART1, C. E. Harris, ed. (1999), 275–283.

Harlow, D. G., and Wei, R. P., "Probabilities of Occurrence and Detection of Damage in Airframe Materials," Fat. & Fract. of Engr. Matls & Structures, 22 (1999), 427–436.

Wei, R. P., Li, C., Harlow, D. G., and Flournoy, T. H. "Probability Modeling of Corrosion Fatigue Crack Growth and Pitting Corrosion," in Fatigue in New and Ageing Aircraft, ICAF 97, Proceedings of the 19th Symposium of the International Committee on Aeronautical Fatigue 18–20 June 1997, Edinburgh, Scotland, Vol. 1 (1997), 197–214.

Wei, R. P., and Harlow, D. G., "Probabilities of Occurrence and Detection, and Airworthiness Assessment," Proceedings of ICAF'99 Symposium on Structural Integrity for the Next Millennium, Bellevue, WA July 12–16, 1999.

Harlow, D. G., and Wei, R. P., "Aging of Airframe Materials: Probability of Occurrence Versus Probability of Detection," 2nd Joint NASA/FAA/DoD Conference on Aging Aircraft,

Williamsburg, VA, 31 August–3 Sept. 1998, NASA/CP-1999-208982/PART1, C. E. Harris, ed. (1999), 275–283.

Wei, R. P., and Harlow, D. G., "Corrosion and Corrosion Fatigue of Aluminum Alloys – An Aging Aircraft Issue," Proceedings of the Seventh International Fatigue Conference (FATIGUE '99), Beijing, China, June 8–12 (1999).

Harlow, D. G., and Wei, R. P., "Probabilities of Occurrence and Detection of Damage in Airframe Materials," Fat. & Fract. of Engr. Matls & Structures, 22 (1999), 427–436.

Wei, R. P., "Corrosion/Corrosion Fatigue and Life-Cycle Management," Mat. Sci. Research International, 7, 3 (2001), 147–156.

Harlow, D. G., and Wei, R. P., "Life Prediction – The Need for a Mechanistically Based Probability Approach," Key Engineering Materials, Trans Tech Publications, Switzerland, 200 (2001), 119–138.

Latham, M., M. C., Harlow, D. G., and Wei, R. P., "Nature and Distribution of Corrosion Fatigue Damage in the Wingskin Fastener Holes of a Boeing 707," "Design for Durability in the Digital Age," Proceedings of the Symposium of the International Committee on Aeronautical Fatigue (ICAF'01), J. Rouchon, Cepadius-Editions, Toulouse, eds., France (2002), 469–484.

Harlow, D. G., and Wei, R. P., "Probability Modelling and Statistical Analysis of Damage in the Lower Wing Skins of Two Retired B-707 Aircraft," Blackwell Science Ltd. Fatigue Fract Engng Mater Struct 24 (2001), 523–535.

Harlow, D. G., and Wei, R. P., "A Critical Comparison between Mechanistically Based Probability and Statistically Based Modeling for Materials Aging," Mater. Sci. & Eng. (2002), 278–284.

Wei, R. P., and Harlow, D. G., "Corrosion-Enhanced Fatigue and Multiple-Site Damage," AIAA Journal, 41, 10 (2003), 2045–2050.

Harlow, D. G., and Wei, R. P., "Linkage Between Safe-Life and Crack Growth Approaches for Fatigue Life Prediction," in Materials Lifetime Science & Engineering, P. K. Liaw, R. A. Buchanan, D. L. Klarstrom, R. P. Wei, D. G. Harlow, and P. F. Tortorelli, eds., The Minerals, Metals & Materials Society, Warrendale, PA (2003).

Wei, R. P., and Harlow, D. G., "Materials Aging and Structural Reliability a Case for Science Based Probability Modeling," ATEM '03, Japan Society of Mechanical Engineers Materials and Mechanics Division, September 10–12 (2003).

Wei, R. P., and Harlow, D. G., "Mechanistically Based Probability Modelling, Life Prediction and Reliability Assessment," Modelling Simul. Mater. Sci. Eng. 13 (2005), R33–R51.

Harlow, D. G., Wei, R. P., Sakai, T., and Oguma, N., "Crack Growth Based Probability Modeling of S-N Response for High Strength Steel," Intl. J. of Fatigue, 28 (2006), 1479–1485.

Harlow, D. G., and Wei, R. P., "Probability Modeling and Material Microstructure Applied to Corrosion and Fatigue of Aluminum and Steel Alloys," Engineering Fracture Mechanics, 76, 5 (2009), 695–708.

FAILURE INVESTIGATIONS/ANALYSES

Wei, R. P., Baker, A. J., Birkle, A. J., and Trozzo, P. S., "Metallographic Examination of Fracture Origin Sites," included as Appendix A in "Investigation of Hydrotest Failure of Thiokol Chemical Corporation 260-Inch-Diameter SL-1 Motor Case," by J. E. Srawley and J. B. Esgar, NASA TMX209;1194 (January 1966).

ANALYTICAL/EXPERIMENTAL TECHNIQUES

Li C.-Y., and Wei, R. P., "Calibrating the Electrical Potential Method for Studying Slow Crack Growth," Materials Research and Standards, ASTM, 6, 8 (1966), 392.

Wei, R. P., Novak, S. R., and Williams, D. P., "Some Important Considerations in the Development of Stress Corrosion Cracking Test Methods," AGARD Conf. Proc. No. 98, *Specialists Meeting on Stress Corrosion Testing Methods* 1971, and Materials Research and Standards, ASTM, 12, 9 (1972), 25.

Wei, R. P., and Brazill, R. L., "An a.c. Potential System for Crack Length Measurement," in *The Measurement of Crack Length and Shape During Fracture and Fatigue*, C. J. Beevers, ed., Engineering Materials Advisory Services Ltd, Warley, UK (1980).

Wei, R. P., and Brazill, R. L., "An Assessment of A-C and D-C Potential Systems for Monitoring Fatigue Crack Growth," in *Fatigue Crack Growth Measurement and Data Analysis*, ASTM STP 738, S. J. Hudak, Jr., and R. J. Bucci, eds., American Society for Testing and Materials, Philadelphia, PA (1981), 103–119.

Alavi, A., Miller, C. D., and Wei, R. P., "A Technique for Measuring the Kinetics of Electrochemical Reactions With Bare Metal Surfaces," Corrosion, 43, 4 (1987), 204–207.

Wei, R. P., and Alavi, A., "A 4-Electrode Analogue for Estimating Electrochemical Reactions with Bare Metal Surfaces at the Crack Tip," Scripta Met., 22 (1988), 969–974.

Wei, R. P., and Alavi, A., "In Situ Techniques for Studying Transient Reactions with Bare Steel Surfaces," J. Electrochem. Soc., 138, 10 (1991), 2907–2912.

Wan, K.-C., Chen, G. S., Gao, M., and Wei, R. P., "Technical Note on The Conventional K Calibration Equations for Single-Edge-Cracked Tension Specimens," Engr. Fract. Mech., 54, 2 (1996), 301–305.

Thomas, J. P., and Wei, R. P., "Standard-Error Estimates for Rates of Change From Indirect Measurements," TECHNOMETRICS, 38, 1 (1996), 59–68.

Wan, K.-C., Chen, G. S., Gao, M., and Wei, R. P., "Technical Note on The Conventional K Calibration Equations for Single-Edge-Cracked Tension Specimens," Engr. Fract. Mech., 54, 2 (1996), 301–305.

Rong, Y., He, G., Chen, S., Hu, G., Gao, M., and Wei, R. P., "On the Methods of Beam Direction and Misorientation Angle/Axis Determination by Systematic Tilt," Journal of Materials Science and Technology, 15, 5 (1999), 410–414.